CAMBRIDGE LIBRARY COLLECTION

Books of enduring scholarly value

Earth Sciences

In the nineteenth century, geology emerged as a distinct academic discipline. It pointed the way towards the theory of evolution, as scientists including Gideon Mantell, Adam Sedgwick, Charles Lyell and Roderick Murchison began to use the evidence of minerals, rock formations and fossils to demonstrate that the earth was older by millions of years than the conventional, Bible-based wisdom had supposed. They argued convincingly that the climate, flora and fauna of the distant past could be deduced from geological evidence. Volcanic activity, the formation of mountains, and the action of glaciers and rivers, tides and ocean currents also became better understood. This series includes landmark publications by pioneers of the modern earth sciences, who advanced the scientific understanding of our planet and the processes by which it is constantly re-shaped.

An Essay towards a Natural History of the Earth

For the physician and natural historian John Woodward (*c.*1655–1728), fossils were the key to unlocking the mystery of the Earth's past, which he attempted to do in this controversial work, first published in 1695 and here reissued in the 1723 third edition. Woodward argues that the 'whole Terrestrial Globe was taken all to Pieces, and dissolved at the Deluge', and that fossilised remains were proof of the flood as described in the Bible. In the first part of the work, Woodward examines other theories of the Earth's history before presenting evidence – much of it based on his own fossil collection – in support of his theory. The work immediately prompted heated debate among his scientific contemporaries. Despite the controversy, Woodward was acknowledged as an expert on fossil classification, cementing this reputation with his influential *Fossils of All Kinds* (1728), which is also reissued in the Cambridge Library Collection.

Cambridge University Press has long been a pioneer in the reissuing of out-of-print titles from its own backlist, producing digital reprints of books that are still sought after by scholars and students but could not be reprinted economically using traditional technology. The Cambridge Library Collection extends this activity to a wider range of books which are still of importance to researchers and professionals, either for the source material they contain, or as landmarks in the history of their academic discipline.

Drawing from the world-renowned collections in the Cambridge University Library and other partner libraries, and guided by the advice of experts in each subject area, Cambridge University Press is using state-of-the-art scanning machines in its own Printing House to capture the content of each book selected for inclusion. The files are processed to give a consistently clear, crisp image, and the books finished to the high quality standard for which the Press is recognised around the world. The latest print-on-demand technology ensures that the books will remain available indefinitely, and that orders for single or multiple copies can quickly be supplied.

The Cambridge Library Collection brings back to life books of enduring scholarly value (including out-of-copyright works originally issued by other publishers) across a wide range of disciplines in the humanities and social sciences and in science and technology.

An Essay towards a Natural History of the Earth

And Terrestrial Bodyes, Especially Minerals

John Woodward

CAMBRIDGE
UNIVERSITY PRESS

University Printing House, Cambridge, CB2 8BS, United Kingdom

Cambridge University Press is part of the University of Cambridge.
It furthers the University's mission by disseminating knowledge in the pursuit of education, learning and research at the highest international levels of excellence.

www.cambridge.org
Information on this title: www.cambridge.org/9781108076982

This edition first published 1723
This digitally printed version 2014

ISBN 978-1-108-07698-2 Paperback

An ESSAY towards a Natural Hiſtory OF THE EARTH, AND *Terreſtrial Bodyes,* ESPECIALY MINERALS:

As alſo of the SEA, RIVERS, and SPRINGS. With an Account of the UNIVERSAL DELUGE: And of the *Effects* that it had upon the EARTH.

By JOHN WOODWARD, M. D. *Profeſſor of Phyſick in* Greſham-College : *Fellow of the* College *of* Phyſicians, *and of the* Royal-Society.

The Third Edition.

LONDON:
Printed for A. BETTESWORTH and W. TAYLOR in *Pater-noſter Row,* R. GOSLING at the *Middle-Temple-Gate* in *Fleet-ſtreet,* and J. CLARKE under the *Royal-Exchange* in *Cornhill,* 1723.

THE PREFACE.

AVING, in the Essay *it self, given some Intimation both of the Design of it, and the Reasons which induced me to make it publick, I shall not here keep the* Reader *in Suspense much longer than while I acquaint him that, proposing to draw a considerable Number of Materials into so narrow a Compass, that they might all be contain'd in this small Volume, I was obliged to be very brief and concise. And therefore, as* Pieces *of Miniature, Sculpture, or other Workmanship in Little, must be allow'd a closer Inspection, so this Treatise*

will require ſome Care and Application in the Peruſal.

*N*OT *but that I have endeavour'd, as far as was practicable in this ſhort Compaſs, ſo to diſpoſe and order* Things*, by interweaving with the Aſſertions ſome of the Proofs whereon they depend, and occaſionaly ſcattering ſeveral of the more important Obſervations throughout the Work, that it will be no very hard Taſk for any one to diſcover the main Grounds whereon all that I here advance is founded.*

*T*HAT *this may be the more clearly apprehended, I ſhall beg Leave to illuſtrate it by one or two Inſtances. It will perhaps, at firſt Sight, ſeem very ſtrange, and almoſt ſhock an ordinary* Reader *to find me aſſerting, as I do, that the whole Terreſtrial Globe was taken all to Pieces, and diſſolved at the Deluge, the Particles of Stone, Marble, and all other ſolid Foſſils being diſſevered, taken up into the Water, and there ſuſtained together with Sea-Shells and other Animal*

nimal and Vegetable Bodyes: and that the present Earth consists, and was formed out of that promiscuous Mass of Sand, Earth, Shells, and the rest, falling down again, and subsiding from the Water. But whoever shall duely attend to what I elsewhere lay down, viz. *That there are vast Multitudes of Shells, and other Marine Bodyes, found at this Day incorporated with and lodged in all Sorts of Stone, in Marble, in Chalk, and to be short, in all the other ordinary Matter of the Globe which is close and compact enough to preserve them: that these are found, thus reposited amongst this Terrestrial Matter, from near the Surface of the Earth downwards to the greatest Depth we ever dig or lay it open, and this in all Parts of it quite round the Globe: that the said Terrestrial Matter is disposed into* Strata *or Layers, placed one upon another, in like Manner as any earthy Sediment, setling down from a Fluid in great Quantity, will naturaly be: that these Marine Bodyes are now found lodged in those* Strata *according to the Order of their*

Gravity, thoſe which are heavyeſt lying deepeſt in the Earth, and the lighter Sorts (when there are any ſuch in the ſame Place) ſhallower or nearer to the Surface: and both thoſe and theſe amongſt Terreſtrial Matter which is of the ſame ſpecifick Gravity that they are, the heavyer Shells in Stone, the lighter in Chalk, and ſo of the reſt; I ſay, whoever ſhall but rightly weigh all this, he'll have no Need to go further for Proof that the Earth was actualy ſo diſſolv'd, and afterwards framed anew, in ſuch Manner as I have ſet forth. And if to this he ſhall think fit to add the other Arguments of the ſame Thing which he will meet with in their Place, they alſo will, I hope, not fail of doing their Part in convincing him ſtill more of the Truth and Certainty of this Matter.

THE other Inſtance I make Choice of ſhall be of the Univerſality of the Deluge, which is another Propoſition that I inſiſt upon. And for this, let but the Reader pleaſe to conſider, what I deliver from authentick Relations, that

that the marine Bodyes aforesaid are found in all Parts of the known World, as well in Europe, Africa, *and* America, *as in* Asia, *and this even to the very Tops of the highest Mountains: and then I think he cannot reasonably doubt of the Proposition; but more especialy if hereunto he shall joyn what I offer concerning the Great Abyss, and thence learn that there is at this Day resident, in that huge Conceptacle, Water enough to effect such a Deluge, to drown the whole Globe, and lay all, even the highest Mountains under Water. But if he should be at a Loss to know how I got such Notice of that subterranean Reservatory as to enable me to make a Computation of the Quantity of Water now conceal'd therein, if he carefully peruse the Propositions concerning Earthquakes, and some others, in the Third Part, he cannot but discover at least some of the Ways whereby I got Light thereinto: and at the same Time find why it is that I am so particular in relating the* Phœnomena *of Earthquakes, and dwell so long upon that Subject in this shorter Work.*

THESE I intend for Example and Direction to the Reader *how he may satisfy himself in any of the other Heads. 'Tis impossible for me to foresee the Difficulties and Hæsitations of every one; they will be more or fewer, according to the Capacity of each Peruser, and as his Penetration and Insight into Nature is greater or less. Those who have Attention enough to take in the intire Platform as here laid down: who see the Chain which runs through the whole: and can pick up and bear in Mind the Observations and Proofs here and there as they lie, and then confer them with the Propositions, will discern, in great Measure, how those Propositions flow from them; but they, who cannot so easyly do this; must be intreated to have a little Patience, untill the Thing be farther unfolded, and more amply and plainly made out.*

A FEW advances there are, in the following Papers, tending to assert the Superintendence and Agency of *of Providence in the natural World: as also to evince the Fidelity and Exactness of the* Mosaic *Narrative of the Creation, and of the Deluge. Which 'tis not improbable but some may be apt to stumble at, and think strange that, in a Physical Discourse, as this is, I should intermeddle with Matters of that Kind. But I may very safely say, that, as to the former, I have not enter'd farther into it than meerly I was lead by the Necessity of my Subject: nor could I have done less than I have, without the most apparent Injury and Injustice to Truth. And for* Moses, *he having given an Account of some Things which I here treat of, I was bound to allow him the same Plea that I do other Writers, and to consider what he hath deliver'd. In order to this I set aside every Thing that might byass my Mind, over-awe, or mislead me in the Scrutiny: and therefore have Regard to him here only as an Historian. I freely bring what he hath related to the Test, comparing it with Things as now they stand: and finding his Account to be punctu-*

aly

aly true, I fairly declare what I find; wherein I do him but ſimply Right, and only the ſame that I would to a common Hiſtorian, to Beroſus *or* Manetho, *to* Herodotus *or* Livy, *on like Occaſion.*

The

The CONTENTS.

PART

The CONTENTS.

PART III.

Concerning the Fluids *of the* Globe.

PART IV.

PART V.

PART VI.

ERRATA.

Page 80, *Line* 5, *for* by *read* ly. *p.* 83, *l.* 12, *f.* requite *r.* requisite. *p.* 163, *in the Margin, to* Pag. *add* 108 supra. *p.* 196, *l.* 20, *f.* Oxyx's *r.* Onyx's. *p.* 243, *l. penult. r.* first out of. *p.* 277, *in the Margin at the Bottom, f.* pag. 25. *r.* pag. 27.

An

An Account of the Obſervations *upon which this Diſcourſe is founded.*

FROM a long Train of *Experience* the World is at length convinced that *Obſervations* are the only ſure *Grounds* whereon to build a laſting and ſubſtantial *Philoſophy*. All *Partyes* are ſo far agreed upon this Matter, that it ſeems to be now the common *Senſe* of *Mankind*.

For which Reaſon I ſhall, in the *Work* before me, give my ſelf up to be guided wholey by *Matter of Fact*; as intending to ſteer that *Courſe* which is thus agreed of all Hands to be the *beſt* and *ſureſt*: and not to offer any thing but what hath due Warrant from *Obſervations*: and thoſe both carefully made, and faithfully related.

And

And that each Reader may the better inform himſelf, not only of *what Sort* my preſent *Obſervations* are, but ſee in what *Manner* alſo, and with what kind of *Accuracy* they were made, 'twill be convenient to give ſome Light into that Matter, and to begin with an *Account* of them; whereby he may be enabled to judge how far they may be *rely'd* upon, and what Meaſure of *Aſſent* the *Propoſitions* which I draw from them may claim.

But before I go any farther, I ought to put in a *Caution*, that an *ample* and *prolix Relation* either of the *Obſervations* themſelves, or of the *Deductions* from them, is not to be expected here. I intend *this* but for a *Scheme* of a *larger Deſign*, and as a *Sample* of what I hope, in good Time, more fully to diſcuſs and make out; propoſing no more, in *this Treatiſe*, than only, in a few plain Words, to deliver my *Sentiments* on certain *Heads* of *Natural Hiſtory*, with ſome of the *Reaſons* and *Grounds* of them, in order to give ſomewhat of preſent Satisfaction to the Curioſity and Demands of ſome of my Friends.

The

The *Obſervations* I ſpeak of were all made in *England*, the far greateſt Part whereof I travell'd over on *purpoſe* to make them; profeſſedly ſearching *all Places* as I paſs'd along, and taking a careful and exact View of *Things* on all Hands as they preſented; in order to inform my ſelf of the *preſent Condition* of the Earth, and all *Bodyes* contain'd in it, as far as either *Grotto's*, or other *Natural Caverns*, or *Mines*, *Quarries*, *Colepits*, and the like, let me into it, and diſplayed to Sight the *interiour Parts* of it. Nor, in the mean Time, did I neglect the *exteriour* or *Surface*: and ſuch *Productions* of it as any where occurr'd, *Plants*, *Inſects*, *Sea*, *River*, and *Land Shells*: and, in a Word, whatever either the *Vegetable* or *Animal World* afforded.

Neither did I confine theſe *Obſervations* to *Land*, or the *Terreſtrial Parts* of the *Globe* only, but extended them to the *Fluids* of it likewiſe, as well thoſe within it, the *Water* of *Mines*, of *Grotto's*, and other ſuch like *Receſſes*, as thoſe upon the *Surface* of it, the *Sea*, *Rivers*, and *Springs*.

My

My principal Intention indeed was to get as compleat and ſatisfactory *Information* of the whole *Mineral Kingdom* as I poſſibly could. To which End I made ſtrict *Enquiry* wherever I came, and laid out for Intelligence of *all Places* where the *Entrails* of the *Earth* were *laid open*, either by Nature, (if I may ſo ſay) or by Art, and humane Induſtry. And whereſoever I had Notice of any conſiderable *natural Spelunca* or *Grotto*: any ſinking of *Wells*: or digging for *Earths*, *Clays*, *Marle*, *Sand*, *Gravel*, *Chalk*, *Cole*, *Stone*, *Marble*, *Ores* of *Metalls*, or the like, I forthwith had recourſe thereunto; where taking a juſt Account of every obſervable *Circumſtance* of the Earth, Stone, Metall, or other Matter, from the *Surface* quite down to the *Bottom* of the *Pit*, I enter'd it carefully into a *Journal*, which I carry'd along with me for that Purpoſe. And ſo paſſing on from Place to Place, I *noted* whatever I found *memorable* in each particular *Pit*, *Quarry*, or *Mine*: and 'tis out of theſe *Notes* that my *Obſervations* are compil'd.

After

After I had *finiſh'd* theſe *Obſervations*, and was *returned* back to this City, ſuch were the *Commotions* which then ſo unhappily diſturbed all *Europe*, that I ſaw I muſt neceſſarily deſiſt here, and ſit down (for the preſent at leaſt) with what I had already done; having little Proſpect of an Opportunity of *carrying on* theſe *Obſervations* any *farther*, or of going *beyond Seas*, to conſider the *State* of the *Earth*, and of all Sorts of *Foſſils*, in more *diſtant Countries*.

But to *ſupply*, as far as poſſible, that *Defect*, I made Application to *Perſons* who had already *travelled*, and I knew were of ſuch *Integrity*, that they would not impoſe *uncertain* or *falſe Relations* upon me: as alſo of ſo much *Curioſity* as to be likely to give me ſome tolerable *Inſight* into the Condition of *theſe Things* in *Foreign Regions*. I likewiſe drew up a *Liſt* of *Quæries* upon this Subject; which I diſpatch'd into all Parts of the *World*, far and near, wherever either I my ſelf, or any of my Acquaintance, had any

Friend resident to transmitt those *Quæries* unto.

The Result was, that in time I was abundantly assured, that the *Circumstances* of *these Things* in *remoter Countries* were much the *same* with those of ours *here*: that the *Stone*, and other *terrestrial Matter*, in *France*, *Flanders*, *Holland*, *Spain*, *Italy*, *Germany*, *Denmark*, *Norway*, and *Sweden*, was distinguish'd into *Strata*, or *Layers*, as it is in *England*: that those *Strata* were divided by *parallel Fissures*: that there were enclosed in the *Stone*, and all the other denser kinds of *terrestrial Matter*, great Numbers of *Shells*, and other Productions of the *Sea*; in the same Manner as in *that* of *this Island*. To be short, by the same Means I got sufficient Intelligence that *these Things* were found in like Manner in *Barbary*, in *Egypt*, in *Guiney*, and other Parts of *Africa*: in *Arabia*, *Syria*, *Persia*, *Malabar*, *China*, and other *Asiatick* Provinces: in *Jamaica*, *Barbadoes*, *Virginia*, *New-England*, *Brasil*, *Peru*, and other Parts of *America*. But I reserve the more *particular*

particular Relation hereof to its proper Place.

So that though my *own Obſervations* were confined to *England*, yet by this Means I was made acquainted with the State of *theſe Bodies* in *other Countries* ; even in almoſt *all Parts* of the *World* wherewith the *Engliſh* maintain any Commerce or Correſpondence : and learn'd, from all Hands, that the *State* of them *there* was conformable to that of ours *here*, in the main, and as far as I ſhall lay any Streſs upon it in my *Concluſions* ; which indeed are not built upon any *Niceties*, or ſolitary and uncommon *Appearances*, but on the moſt ſimple and obvious *Circumſtances* of theſe terreſtrial *Bodies*.

As to the *Certainty* and *Accurateneſs* of my *Obſervations*, thus much may modeſtly and very truly be ſaid, that I do not offer *any one* before I had firſt thoroughly and clearly *informed* my ſelf in all material *Circumſtances* of it : and had Opportunity of obſerving it in *more Places* than *one*, that I might be ſatisfy'd there was nothing *caſual* or *contingent* in any of thoſe Circumſtances. This will not be thought an over-great *Exactneſs*, or any thing more than was *needful*, by *thoſe* who have noted how much *Philoſophy* hath *ſuffered* by the *Neglect* and *Overſight* of ſome *Naturaliſts* in this Reſpect. A *tranſient* and *perfunctory Examination* of Things, frequently leads Men into conſiderable *Miſtakes*, which a more *correct* and *rigorous Scrutiny* would have detected and avoided. The Truth is, I have been the more ſcrupulous and wary in regard the *Inferences* drawn from theſe *Obſervations* are of ſome *Importance*. 'Twas but neceſſary that the *Foundation* ſhould be *firm*, when a *Superſtructure* of *Bulk* and *Weight* was to be rais'd upon it. And therefore I advance nothing from any Obſervation that was not made with this *Caution*, and that any Man may not, as well as my ſelf, without any great Pains, inform himſelf of the Truth of. Now, as long as the next Cole-pit, or Mine, the next Quarry, or Chalk-pit, will give abundant *Atteſtation* to what I write *theſe* are ſo ready and

and obvious in almoſt all Places, that I need not be any where far to ſeek for a *Compurgator:* and to theſe I may very ſafely appeal.

Concerning the *Obſervations* themſelves therefore, there cannot well ariſe any *Doubt* but what may eaſily be ſatisfy'd: and what I propoſe in this *Eſſay* being *founded* upon theſe *Obſervations*, every *Reader* will be Judge of the *Truth* and *Probability* of it, and whether *that* which I do ſo propoſe naturaly *follows* from them or not.

I ſhall diſtribute *them* into *two* general *Claſſes* or *Sections*, whereof the *former* will comprehend my Obſervations upon all the *Terreſtrial Matter* that is naturaly *diſpoſed* into *Layers*, or *Strata*; ſuch as our *common Sand-Stone*, *Marble*, *Cole*, *Chalk*, all Sorts of *Earth*, *Marle*, *Clay*, *Sand*, with ſome others.

Of this various Matter, thus formed into *Strata*, the far *greateſt Part* of the *Terreſtrial Globe* conſiſts, from its *Surface* downwards to the *greateſt Depth* we ever dig or mine. And it is upon my Obſervations on *this* that I have grounded all my *general*

Concluſions concerning the *Earth*: all that relate to its Form: all that relate to the *Univerſal* and *other Deluges*: in a Word, all that relate to the ſeveral *Viciſſitudes* and *Alterations* that it hath yet undergone. Nay, upon the ſame Obſervations I have alſo founded ſeveral *Concluſions* touching *Metalls*, *Spar*, and other *Minerals*, which are found lodged either in theſe *Strata*, amongſt the Sand, Chalk, Earth, and the reſt: or in the *perpendicular Intervalls* of the *Strata* of Stone, Marble, or other ſolid Matter.

For upon the *particular* Obſervations on the ſaid *Metallick* and *Mineral Bodies*, (which are the Subjects of the *ſecond Section*,) I have not founded any thing but what purely and immediately concerns the *Natural Hiſtory* of *thoſe Bodies*.

To proceed therefore to the Account of my Obſervations upon *Sand ſtone*. And in *theſe*, though I do not neglect to note the *ſeveral Kinds* or Varieties of it: Free-ſtone, Ragg-ſtone, Lime-ſtone, and the reſt: the different *Hardneſs*, or *Solidity*, of each: as alſo its *Colour*, *Texture*,

Texture, and the *peculiar Matter* which conſtitutes it; yet I confine my ſelf more ſtrictly to conſider the *Manner* how 'tis *diſpoſed* in the *Earth*: the *Strata*, into which, by means of *horizontal* * and parallel *Fiſſures*, it is divided: the *Order* and *Number* of theſe *Strata*: their *Situation* in reſpect of the Horizon: the *Thickneſs*, *Depth*, and other Circumſtances of each: the *Interruptions* of the *Strata*, I mean the *perpendicular* * *Fiſſures*, which interſect the *horizontal* ones: the different *Capacity* or *Largeneſs* of theſe *perpendicular Intervalls*: their *Diſtances* from each other: and the *Spar*, and other *Mineral* and *Metallick Matter*, uſualy contained in them.

But, becauſe I ſaw that *Deductions* of conſiderable *Import* and *Conſequence* might be drawn from them, I have with great *Care* and *Intention* obſerved the *Condition* of ſuch *hete-*

* * I call thoſe *Fiſſures*, which diſtinguiſh the Stone into *Strata*, *Horizontal* ones: and thoſe which interſect theſe, *Perpendicular*; not ſo much with reſpect to the *preſent* Site of the *Strata*, which is alter'd, in many Places, and now much different from their *original* Situation, concerning which, ſee *Part* 2. *Conſect.* 5, & 6.

rogeneous Bodies, which I found immerſed and included in the *Maſs* of this *Sand-ſtone*; particularly the *Shells* of *Oyſters*, *Muſcles*, *Scallopes*, *Cockles*, *Periwincles*, and very many other *marine Productions*. I have, I ſay, very diligently noted all *Circumſtances* of theſe *Shells*: the vaſt *Numbers* of them: the ſeveral *Kinds* that are thus lodged in the Subſtance of the Stone: the *Order* and *Manner* of their Poſition in it: the ſeveral *Depths* at which they are found: the *Matter* which they *contain* in them, and wherewith their *Cavities* are uſualy filled.

Theſe *Obſervations* about Stone are ſucceeded by *others*, of like Nature, concerning *Marble*, *Cole*, and *Chalk*: their *Fiſſures*: the *Situation* of their *Strata*: the *Shells*, and other *heterogeneous Bodies* lodged therein.

In the next Place, *thoſe* which concern *Marle*, *Clay*, the ſeveral Kinds of *Earth*, *Sand*, and ſome other *Foſſils*: the *Shells* and other like *Bodies*, lodged in their *Strata*: the *Poſition* of thoſe *Stata*: their *Order*: their *Diſtinctions* from each other, by the Difference of the

Matter

Matter of each, and by its different *Consistence* and *Colour*; the *Strata* of these *laxer* Kinds of *Matter* being not ordinarily *divided* from each other by Interposition of *horizontal Fissures*, as those of Stone, and such other *solid Matter*, constantly are.

And lastly, those which relate to the upper or outmost *Stratum* of all: I mean that *blackish Layer* of *Earth* or *Mould* which is called by some *Garden-Earth*, by others *Under-turf-Earth*, wherewith the *Terrestrial Globe* is almost every where *invested*, unless it be disturbed, or flung off by Rains, Digging, Plowing, or some other external Force; insomuch, that whatsoever lies deeper, or underneath, whether Stone, Marble, Chalk, Gravel, or whatever else, *this Stratum* is still expanded at Top of all; serving, as it were, for a *common Integument* to the rest: and being (as shall be shewn in due Place*) the *Seminary* or *Promptuary* that furnisheth forth *Matter* for the Formation and Increment of *Animal* and *Vegetable Bodies*;

* Part. V. Consect. 1.

Bodies; and into which all of them ſucceſſively are again finaly *returned.*

The *Obſervations* being thus diſpatch'd, my next Step ſhould have been to have propoſed the *Deductions* from them: to have determin'd how theſe *Sea-Shells* were brought to *Land*, and how they became interr'd in the Bowels of the *Earth*, in the Manner deſcribed in thoſe Obſervations. But, before I could proceed any farther towards *that*, I found my ſelf neceſſarily obliged to take off a *Difficulty* ſtarted by ſome *learned Men* who have wrote now *lately* upon the Subject, and aſſert that theſe *Shells* are not *real*: that they were never bred at *Sea*: but are all of *Terreſtrial Original*, being *meer Stones*, though they bear a Reſemblance of *Shells*, and formed, in the *Places* where they are *now found*, by a kind of *Luſus* of *Nature*, in Imitation of *Shells.*

How nearly I am concerned to *remove* this *Obſtacle*, before I paſs on any farther to the Proſecution of my *Deſign*, any one may preſently ſee. For to go about to enquire at *what Time*, and by *what Means*

Means these *Shells* were conveyed out of the *Sea* to dry *Land*, when a *Doubt* hath been moved whether they are *Shells* or not, or ever belonged to the *Sea*, without first *clearing* this *Matter*, and putting it quite out of Doubt, would be very absurd. In order therefore unto *this*, I premise,

A Disser-

A Dissertation *concerning* Shells, *and other* marine Bodies, *found at* Land; *Proving that they were originaly generated and formed at* Sea: *that they are the real Spoils of once* living Animals: *and not* Stones, *or natural* Fossils, *as some late Learned Men have thought.*

IN my Extract of this *Dissertation* I shall fairly, and in as little Compass as may be, lay before the *Reader*, first the *Arguments* that have been urged by *those Writers* to perswade us that these *Bodies* are meer *Mineral Substances*. And having detected the *Insufficiency* of them, by evincing how far they are from being *conclusive*, and how much they fall short of *proving* what they are alledged for, I shall then proceed to lay

lay down a brief Scheme of *my own*, and offer ſome of the *Reaſons* which have induced me to believe that theſe are the very *Exuviæ* of *Animals*, and all owing to the *Sea*.

I would not be thought to inſinuate that the *Opinion* of *thoſe Gentlemen* carries no Shew of *Truth*, nor Umbrage of *Reaſon* of its Side. 'Tis not to be ſuppoſed, that *Perſons* of their *Learning* and *Abilities* would ever have eſpouſed it, were it not in ſome Meaſure *plauſible*: and had not at leaſt a fair Appearance of *Probability*. The very finding theſe *Bodies* included in *Stone*, and lodged in the *Earth* together with *Minerals*, was alone enough to move a Suſpicion that *theſe* were *Minerals* too. The finding them even to the very *Bottom* of *Quarries* and *Mines*: in the moſt retired and *inward Parts* of the moſt *firm* and *ſolid Rocks*: in the *deepeſt* Bowels of the *Earth*, as well as upon the Surface of it: upon the *Tops* of even the *higheſt Hills* and *Mountains*, as well as in the *Valleys* and *Plains*: and this not in this or that *Province* only, not only in one or two *Fields*, but

but almoſt *every where:* in *all Countries* and Quarters of the *Globe*, wherever there is any *digging* for Marble, for Stone, for Chalk, or any other Terreſtrial Matter that is ſo *compact* as to fence off *external Injuries*, and ſhield them from Decay and Rottenneſs. *This*, toge-
1. " ther with their being lodged in " company of the Belemnites, Selenites, Marchaſits, Flints, and " other like *Bodies*, which were " inconteſtibly *natural Foſſils*, and, " as *they* ſuppoſed, in the *Place* of " their *Formation*," was enough to ſtagger a Spectator, and make him ready to entertain a Belief that *theſe were ſo* too. 'Tis a *Phænomenon* ſo ſurprizing and extraordinary, that 'tis not ſtrange that a Man ſhould ſcarcely credit his very *Senſes* in the Caſe: that he ſhould more readily incline to believe that they were *Minerals* as the Belemnites, and the others recited, are: or indeed almoſt any thing elſe rather than *Sea Shells*; eſpecially in ſuch *Multitudes*, and in *Places* ſo unlikely: ſo *deep* in the *Earth*, and *far* from the *Sea*, as theſe are commonly found. Nor

Nor was this, as indeed they tell us, the only *Difficulty* these worthy Persons had to surmount; "They "found, together with *these*, certain *Bodies* that bore the *Shape* "and resemblance of Cockles, "Muscles, and other *Shells*, which "yet were *not realy such*; but consisted intirely, some of them, of "*Sand-stone*: others of *Flint*: and "others of *Spar*: or some other "kind of *Mineral Matter*." 2.

Nay, they met with some, "That "were in all Appearance *Shells*: "that were of the same *Bigness*, "*Figure*, and *Texture*, with the "common Echini, Scallops, and "Perewinkles; but had notwith-"standing *Flint*, *Native-Vitriol*, *Spar*, "*Iron-Ore*, or other *Metallick* or "*Mineral Matter*, either adhering "firmly in Lumps to the *Outsides* "of them, or insinuated into their "*Substance*, into their *Pores*, and "*inner Parts*, so as to disguise them "very much, and give them a *Face* "and *Mien* extremely unlike to "that of those Shells which are at "*this Day* found at *Sea*." 3.

They

4. They obſerv'd alſo, that "a-
" mongſt the *Shells*, that were *fair*,
" *unaltered*, and *free* from ſuch *Mi-*
" *neral Inſinuations*, there were *ſome*
" which could not be *match'd* by
" any Species of Shell-fiſh *now*
" *found* upon the *Sea-Shores*.

5. And that on the contrary, " there
" were ſeveral *Shells* found com-
" monly upon the ſaid *Shores*, ſuch
" as the larger Shells of the *Bucci-*
" *na*, of the *Conchæ Veneris*: of
" *Crabs*, *Lobſters*, and others, both
" of the *Cruſtaceous* and *Teſtaceous*
" *Kinds*, which yet we never meet
" with *at Land*, or in our *Quarries*."

Nay there were ſome other *Diffi-culties* which they have urged, and which (though they be of *leſſer Weight*) ſhall all of them be recounted and conſidered more particularly in due Place.

Upon the whole therefore 'tis very plain, that theſe Authors did not eſpouſe *this Opinion* without ſome *Grounds*, without ſome countenance of *Probability*: and that they have charged the *oppoſite* with a large crowd of *Difficulties*. Yea ſo far are they from being deſtitute of of an handſome *Apology*, that they very well deſerve the *Thanks* of the *World* for what they have done. For, although they have not ſucceeded in their *Attempts* about the *Origin* of *theſe Bodies*, they have made *Diſcoveries* in other Reſpects concerning *them*, and in *other Parts* of *Nature* likewiſe, of that *Moment* and *Conſequence*, as to have thereby laid a great and laſting *Obligation* upon the intelligent and diſcerning Part of *Mankind*.

But that they have failed notwithſtanding in *this Enterprize*, 'tis, I think, not over difficult to *prove*. And *this* is the Subject of the preſent *Diſcourſe*. Wherein I hope to I.
make out, that the *Sea* gave Birth to *theſe Bodies* *: that *they* are ſo far from being formed in the *Earth*, or in the *Places* where they are *now found*, that even the Belemnites, Selenites, Marchaſits, Flints, and other natural *Minerals*, which are lodged in the *Earth*, together with *theſe Shells*, were not formed *there*, but had *Being* before ever they came *thither* : and were fully *formed* and

* *Vid. Pag.* 23, 24, *&c. infra.*

 finiſhed

finiſhed before they were repoſed in that Manner.†

† Confer. Part. IV, Conſect. 2.

2. That the above mentioned *Bodies* which conſiſt of *Stone*, of *Spar*, *Flint*, and the like, and yet carry a *Reſemblance* of Muſcles, Cockles, and other *Shells*, were originaly formed in the *Cavities* of *Shells* of thoſe Kinds which they ſo reſemble; theſe Shells having ſerved as *Matrices* or *Moulds* to them; the Sand, Sparry, and Flinty Matter being then *ſoft*, or in a *State* of *Solution*, and ſo, ſuſceptible of any *Form*, when it was thus introduced into theſe *Shelly-Moulds*: and that it conſolidated, or became *hard* afterwards*.

* Concerning theſe Myitæ, Cochlitæ, &c. See Part 4. Conſ. 2. & Part 5. Conſ. 5. infra.

3. That for the *Metallick* and *Mineral Matter* which ſometimes adheres to the *Surfaces* of theſe *Shells*, or is intruded into their *Pores*, and lodged in the *Interſtices* of their *Fibres*, 'tis all manifeſtly *adventitious* †; the *Mineral Particles* being plainly to be diſtinguiſhed from the *teſtaceous* ones, or the Texture and Subſtance of the *Shell*, by good *Glaſſes*, if not by the naked *Eye*. That though the Thing had been ſo that this *Accretion* had not been thus diſcernible, and

† Part 4. Conſ. 2. infra.

and conſequently the *Alteration* of theſe *Shells* could not have been accounted for, ſo that we had been perfectly in the dark as to the Origin of the Bodies thus *alter'd*, and that nothing at all could have been determined concerning *them*; yet *this* would not have been any the leaſt *Impediment* or *Objection* againſt that which I inſiſt upon; there being ſo *very few* of *theſe* in Compariſon of *thoſe* which have undergone no ſuch *Alteration*. There being, I ſay, beſides *theſe*, ſuch vaſt *Multitudes* of *Shells* contained in *Stone*, &c. which are *intire*, *fair*, and abſolutely free from any ſuch *Mineral Contagion*: which are to be *match'd* by others at this Day found upon our *Shores*, and which do not *differ* in any Reſpect from them; being of the *ſame Size* that thoſe are of, and the *ſame Shape* preciſely: of the *ſame Subſtance* and *Texture*; as conſiſting of the *ſame peculiar Matter*, and this conſtituted and diſpoſed in the *ſame Manner*, as is that of their reſpective Fellow-kinds at *Sea*: the *Tendency* of the *Fibres* and *Striæ* the ſame: the *Compoſition* of the *Lamellæ*, conſtituted by theſe *Fibres*, alike in both : the ſame *Veſtigia* of *Tendons* (by Means whereof the Animal is faſtned and joyned to the *Shell*) in each : the ſame *Papillæ* : the ſame *Sutures*, and every thing elſe, whether *within* or *without* the *Shell*, in its *Cavity*, or upon its *Convexity*, in the *Subſtance*, or upon the *Surface* of it. Beſides, theſe *Foſſil Shells* are attended with the ordinary *Accidents* of the *marine ones*, *ex. gr.* they ſometimes *grow* to one another, the *leſſer Shells* being fixed to the *larger* : they have *Balani*, *Tubuli vermiculares*, *Pearls*, and the like, *ſtill* actualy *growing* upon them. And, which is very conſiderable, they are moſt exactly of the ſame *ſpecifick Gravity* with their Fellow-kinds now upon the *Shores*. Nay farther, they anſwer all *Chymical Tryals* in like Manner as the *Sea-Shells* do : their *Parts* when *diſſolved* have the ſame *Appearance* to View, the ſame *Smell* and *Taſte* : they have the ſame *Vires* and *Effects* in *Medicine*, when inwardly adminiſtred to Animal Bodies : *Aqua fortis*, Oyl of *Vitriol*, and other like *Menſtrua*,

Menſtrua, have the very ſame *Effects* upon both. In a Word, ſo exactly conformable to the *marine ones* are theſe Shells, Teeth, and Bones, which are *digged* up out of the *Earth*, that though ſeveral *Hundreds* of them, which I *now keep* by me, have been nicely and criticaly *examined* by very many *Learned Men*, who are ſkill'd in all Parts of *Natural Hiſtory*, and who have been particularly curious in, and converſant with *Shells*, and other *marine Productions*, yet never any Man of them went away *diſſatisfy'd*, or *doubting* whether theſe are realy the very *Exuviæ* of Sea-fiſhes or not. Nay, which is much more to my Purpoſe, ſome of the moſt *eminent* of thoſe very *Gentlemen*, who formerly were *doubtful* in this *Matter*, and rather inclinable to believe that theſe were *natural Minerals*, and who had *wrote* in Defenſe of that *Opinion*, do, notwithſtanding, upon ſtrict and repeated *Inſpection* of theſe *Bodies* in my *Collection*, and upon *farther Enquiry*, and Procuration of *plain* and *unalter'd* Shells from ſeveral Parts of this *Iſland*, fully *aſ-*

ſent to me herein, and are now *convinced* that theſe áre the real *Spoils* and *Remains* of *Sea-Animals.* And, being thus ſatisfy'd, ſuch is their *Ingenuity*, and ſo great their Affection to *Truth.* that they have perſonaly requeſted me to *publiſh* my *Thoughts* in order to the fuller *clearing* of this *Matter.* But, to proceed,

4. That although I can *pair*, with *Sea-Shells*, ſeveral of theſe *Foſſil ones* that thoſe *Gentlemen* have pronounced altogether *unlike* any thing that the *Salt-Water* produceth, yet 'tis indeed very true that there are found *ſome Shells*, at *Land*, in *Stone*, and in *Chalk*, which cannot probably be *match'd* by any Species of Shells *now* appearing upon our *Shores.* But, notwithſtanding this, I cannot but affirm that *theſe*, even the moſt ſtrange and enormous of them, have all the eſſential Notes and Characters of *Sea-Shells*, and ſhew as near a *Relation* to ſome *now extant* upon the *Shores* as the different Species of *thoſe* themſelves do to one another: that they are of the very ſame *ſpecifick Gravity* with thoſe to which they are ſo generically

caly allied: and of the ſame *Texture* and *Conſtitution* of Parts; the *Subſtance* of *theſe* being as plainly *Teſtaceous*, as *that* of *thoſe* is; inſomuch that any Man that compares them, can no more *doubt* of the *Reality* of the *one* than of the *other*. Whence it muſt needs follow, that there were ſuch Shell-fiſh *once in Being*; which is enough for my Purpoſe; I being no ways concerned to make out that there are of the ſame Kinds *ſtill* actualy *living* in the *Ocean*. Though if I *was*, 'twould be no very hard Taſk; it being evident from the *Relations* of *Dyvers*, and *Fiſhers* for *Pearls*, that there are *many Kinds* of *Shell-fiſh* which lye perpetualy *concealed* in the *Deep*, ſkreen'd from our Eyes by that vaſt World of *Water*, and which have their continual *Abode* at the *Bottom* of the *Ocean*, without ever approaching *near* the *Shores*; it being as unnatural for *theſe* to deſert this their *native Station*, as 'tis for *thoſe* that are the *Inhabitants* of the *Shores* to quit *theirs*, and retire into the *Deep*. For this Reaſon *theſe* are called by Naturaliſts ἐμβύθιοι, and

Pelagiæ: as the *others*, that reside nearer to the Shores, are by them called *Littorales*. Now the Shells which we find *exposed* upon our *Shores*, are only those which are cast up and stranded by *Tides* and by *Storms*: and consequently are all of them *Exuviæ* of those *Kinds* that *live* near the *Shores*, and not of *those* that inhabit the *Main*, or the *deeper* and *remoter* Parts of the *Ocean*; it being certain from the *Relations* also of *Dyvers*, that the *Tides* and *Storms*, even the most tempestuous and turbulent, affect only the *superficial* Parts of the *Ocean*, the *Shallows*, and *Shores*, but never reach the greater *Depths*, or disturb the *Bottom* of the *Main*. *These* are as *quiet*, and free from *Commotion* in the midst of *Storms*, as in the greatest *Calm*. So that the *Shell-fish*, which are resident in *these* Places, live and dye *there*, and are *never* dislodg'd or *removed* by Storms, nor cast upon the *Shores*, which the *Littorales* usualy are. When therefore I shall have proved more at large, that *those* which we find at *Land*, that are not matchable with any

upon

upon our *Shores*, are many of them of *those very Kinds* which the forecited *Relations* particularly assure us are found no were but in the *deeper Parts* of the *Sea*: and that as well *those* which we *can match*, as those we *cannot*, are all *Remains* of the *Universal Deluge*, when the *Water* of the *Ocean*, being boisterously turned *out* upon the *Earth*, bore along with it *Fishes* of all Sorts, *Shells*, and the like moveable Bodies, which it *left behind* at its *Return* back again to its *Chanel*; it will not, I presume, be thought *strange*, that, amongst the *rest*, it left some of the *Pelagiæ*, or those Kinds of *Shells* which naturaly have their Abode at *Main-Sea*, and which therefore are now never flung up upon the *Shores*. And it may very reasonably be concluded, that all these *strange* Shells, which we cannot so *match*, are of these *Pelagiæ*: that the several Kinds of them are at *this Day* living in the *huge Bosom* of the *Ocean*: and that there is not any one intire *Species* of *Shell-fish*, formerly in Being, now *perish'd*, or *lost*.

That

5. That it is alſo very true that there are *ſome Shells*, ſuch as thoſe of the larger *Buccina*, and *Conchæ Veneris*, of *Lobſters*, *Crabs*, and others of the *cruſtaceous Kind*, that are very rarely found at Land; ſo *rarely*, that ſome of theſe Gentlemen have aſſerted that they are *never* found; but *that* I ſhall ſhew to be a *Miſtake*, all the Shells in their whole *Liſt* having been found in the *Earth* in one Place or other. But that theſe are very *ſeldom* found any where, I moſt readily grant; and this is ſo far from being an Argument againſt what I am going to advance, that 'tis as full and ſubſtantial a *Proof* of the *Truth* of it as I could poſſibly wiſh. For the Shells in *this Liſt* are all *lighter* than *Stone*, *Marble*, and the other ordinary *Terreſtrial Matter*. Now both *theſe*, and all other Sorts of *Shells* that are *ſo light*, occurr very *ſeldom* at *Land*, or in the *Earth*, in compariſon of the *Shells* of Cockles, Perewinkles, and the reſt which are more *ponderous*, ſo as to equal the Stone, and the other Terreſtrial Matter in *Gravity*. The Reaſon of which

which will be very plain, when I ſhall have ſhewn * that at the time of the *Deluge* (when theſe Shells were brought out upon the Earth, and repoſed therein in the Manner we now find them) *Stone*, and all other ſolid *Minerals* loſt their *Solidity:* and that the ſever'd Particles thereof, together with thoſe of the Earth, Chalk, and the reſt, as alſo Shells, and all other *Animal* and *Vegetable Bodies*, were *taken up* into, and *ſuſtained in*, the *Water:* that at length all theſe *ſubſided* † again promiſcuouſly, and without any other *Order* than that of the *different Gravity* of the ſeveral *Bodies* in this confuſed *Maſs*; thoſe which had the *greateſt* degree of *Gravity* ſinking down *firſt*, and ſo ſettling *loweſt*; then thoſe Bodies which had a *leſſer Share* of it fell *next*, and ſettled ſo as to make a *Stratum* upon the *former*; and ſo on, in their *ſeveral Turns*, to the *lighteſt* of all, which ſubſiding *laſt*, ſettled at the *Surface*, and cover'd all the reſt. Now this very various *Miſcellany* of *Bodies* being determined to *Subſidence* in this *Order* meerly by their *different Gravities*,

* *Vid.* Part 2. *Conſect.* 2.

† *Vid.* Part 2. *Conſect.* 3.

ties, all those which had the *same degree* of *Gravity* subsided at the *same Time*, fell into, and composed the *same Stratum*. So that those *Shells* and other *Bodies*, that were of the *same Specifick Gravity* with *Sand*, sunk down together with it, and so became inclos'd in the *Strata* of *Stone*, which *that Sand* form'd or constituted. Whilst those which were *lighter*, and of but the *same specifick Gravity* with *Chalk* (in such *Places* of the *Mass* where any *Chalk* was) fell to the Bottom at the *same Time* that the *Chalky Particles* did, and so were entombed in the *Strata* of *Chalk*; and in like Manner all the rest. Accordingly we *now* find in the *Sand-stone* of all Countries (the *specifick Gravity* of the several Sorts whereof is very little different, being generaly to *Water* as $2\frac{1}{2}$ or $2\frac{9}{10}$ to 1) only those *Conchæ*, *Pectines*, *Cochleæ*, and other *Shells* that are nearly of the same *Gravity*, *viz.* $2\frac{1}{2}$ or $2\frac{5}{8}$ to 1. But these are ordinarily found enclos'd in it in *prodigious Numbers*, whereas of *Oyster-shells*, (which are in Gravity but as about

about 2 $\frac{1}{3}$ to 1) of *Echini* (which are but as 2, or 2 $\frac{1}{8}$ to 1) or the other *lighter Kinds* of Shells, *ſcarce one* ever appears therein. On the contrary, in *Chalk* (which is *lighter* than *Stone*, being but as about 2 $\frac{1}{10}$ to 1) there are *only* found *Echini*, and the other *lighter* Sorts of Shells; it being extremely unuſual to meet with ſo much as *one ſingle Shell* of any of all the *heavier Kinds* amongſt *Chalk*; but for the ſaid *Echini*, and other the *lighter* Sorts of *Shells*, they are very *numerous* and *frequent*, in all the *Chalk-pits* of *Kent*, *Surrey*, *Eſſex*, *Hartfordſhire*, *Barkſhire*, *Oxfordſhire*, *Wiltſhire*, and all others that I have ſearch'd; being found indifferently in the *Beds* of *Chalk* from the *Top* quite down to the *Bottom* of the *Pit*; I having my ſelf commonly obſerved them to the very *Bottom* of all, in *Pits* that were an *hundred foot deep*, and in *Wells* much *deeper*. To conclude, thoſe *Shells*, and other *Bodies*, that were *ſtill lighter* than *theſe*, and conſequently *lighter* than *Stone*, *Chalk*, and the other *common Matter* of the *Earth*, ſuch as the Shells of

Lobſters

Lobſters (which are but as 1 $\frac{1}{5}$ to 1) of *Crabs*, (1 $\frac{3}{4}$ to 1) and the reſt of the *Cruſtaceous Kind:* the *Teeth* and *Bones* of the *cartilaginous* and *ſquammoſe Fiſhes*, and many other *Bodies*, theſe I ſay would of Courſe *ſubſide laſt* of all, and ſo, falling above the reſt, be lodged *near*, if not *upon* the *Surface*; where, being continualy expoſed to *Weather*, and other *Injuries*, they muſt in tract of Time needs *decay* and *rot*, and at laſt quite *vaniſh* and *diſappear*; and 'tis not to me any great *Wonder*, that at *this Diſtance* of four thouſand Years, we find ſo *very few* of them remaining. So that I think I may now ſafely appeal to any ingenuous and impartial Looker on, whether this, [*That we find all thoſe Kinds of Shells (now extant upon our Shores) which have nearly the ſame Gravity with Stone, and the other ordinary Matter of our Earth, that is ſo tight and compact as to preſerve them, encloſed in great Plenty therein, and only thoſe, the reſt which are lighter being ſo very rarely found,*] can reaſonably be ſuppoſed to have happened by *meer Chance*, with this *Conſtancy*

Conſtancy and *Certainty*, and that in *ſo many* and *diſtant Places :* as alſo, whether *this* be any *Objection* againſt my *Hypotheſis :* or rather be not the *ſtrongeſt* acceſſary *Confirmation* of it that could well be expected, or even deſir'd.

To this *Diſſertation* I ſhall ſubjoin an *Appendix*, which will conſiſt of *ſeveral Sections*, touching the *Bodies* call'd *Unicornu Foſſile*, *Lapis Judaicus*, *Entrochus*, *Aſteria*, or the Star-ſtone-Columns : with ſome *farther Reflections* upon the *Bufonites*, *Gloſſopetra*, and *Cornu Ammonis* ; proving that *theſe*, and *ſeveral more*, which have been (for many Ages) reputed *Gemms*, and meer *Stones*, are realy nothing elſe but the *Teeth*, *Bones*, and other Parts of *Sea-Animals*, and (as the reſt were) left behind by the *Univerſal Deluge.*

PART

PART I.

An Examination of the Opinions of former Writers *on this Subject. The* Means *whereby* they *thought these* Marine Bodies *brought out upon the* Earth. *Of certain* Changes *of* Sea *and* Land, *and other* Alterations *in the Terraqueous* Globe, *which* they *suppose to have happen'd.*

THIS so considerable a Point being thus gained: the *Legitimacy* or *Reality* of these *Marine Bodies* vindicated and asserted: and my Way so far effectualy cleared by the foregoing *Dissertation*; I now re-assume my *original Design*,

Deſign, and paſs on to inquire by what *Means* they were hurryed out of the *Ocean*, the Place of their *native Abode*, to *dry Land*, and even to *Countries* very *remote* from any *Seas*.

It is indeed a *Queſtion* of great *Antiquity* : and which hath, for *many Ages*, given no ſmall *Fatigue* to *Learned Men*. Nor hath the *preſent* been leſs inquiſitive into this *Affair* than the *former Ages* were. We have ſeen ſeveral Hands imployed herein : and many of them very excellent ones too. The great Number of the *Undertakers*, the *Worth* of ſome of them, and their *Zeal* to bring the Matter to a Deciſion, are ſure *Arguments* of the *Dignity* and *Importance* of it : and, that it is not hitherto decided, is as certain a Proof of its *Difficulty*.

Some were of Opinion that theſe *Shells* were fetch'd from *Sea* by the *ancient Inhabitants* of thoſe Countries where they are now found ; who, after they had uſed the included *Fiſhes* for Food, *flinging forth* the *Shells*, many of them became *petrified*, as they ſpeak ; being thereby preſerved down to our Times, and are the ſame which we at this Day find in our *Fields* and *Quarryes.*

Others rather thought that they were only *Reliques* of ſome former great *Inundations* of the *Sea* ; which, furiouſly ruſhing forth, and *overflowing* the adjacent Territories, bore theſe *Bodies* out upon the *Earth* along with it : but returning at length more leiſurely and calmly back again, it *left* them all *behind.*

Many were of Opinion, that the *Sea* frequently flitted and *changed* its Place : that ſeveral Parts of the *Globe* which are now *dry Land*, and habitable, lay heretofore at the *Bottom* of the *Sea*, and were covered by it : that particularly the very *Countries*, which preſent us with theſe *Spoils* of it, were *anciently* in its *Poſſeſſion* ; being *then* an Habitation of *Sharks* and other *Fiſhes*, of Oyſters, Cockles, and the like ; but the *Sea*, in tract of Time, *retreating* thence, and betaking it ſelf into *new Quarters* : gaining as much Ground on the *oppoſite Coaſts*, as it loſt upon *thoſe*, left theſe *Shells* there

there as Marks of its *ancient Bounds* and *Seat.*

Amongst the rest there were indeed some who believed *these* to be *Remains* of the *General Deluge:* and so many Monuments of that calamitous and fatal *Irruption.* These last assuredly were in the *right*; but the far greater Part of them rather *asserted* than *proved* this: rather deliver'd it as their *Opinion*, than offer'd any *rational Arguments* to induce *others* to the same *Belief.* And for the *rest*, who did offer *any*, so *unhappy* were they in the *Choice*, and *unsuccessful* in the *Management* of them, by reason of the *Shortness* of their *Observations*, and their not having duely informed themselves of the *State* of these *Things*, that none of the other *Partizans* appear'd with less *Applause*, none less strenuously maintain'd their *Ground*, than *these* did.

The Truth is, as Matters were order'd amongst them, no Man could receive much *Light* or *Satisfaction* from what was advanced by any of them. They little more than clash'd with one another. Each could de-

 molish

moliſh the others *Work* with Eaſe enough, but not a Man of them tolerably defend his *own* ; which was ſure never to outſtand the firſt Aſſault that was made. Yea upon ſo equal *Terms* did they all ſtand, that no one could well lay claim to a larger Share of *Truth* for his Side: no one had a fairer Pretence of *Right*, than the reſt. And, it being impoſſible to imagine that *all* could be in the *Right*, ſome *Learned Men* began to ſuſpect that *none* of them were ſo.

Theſe thereupon laid out on all Hands for ſome new *Expedient* to ſolve and put an End to the *Perplexity*. And 'twas this *laſt Effort* that brought forth the *Opinion*, that theſe *Bodies* are not what they ſeem to be : that they are *no Shells*, but meer *Sportings* of active *Nature* in this *ſubterraneous Kingdom*, and only Semblances or Imitations of Shells. They imagined that this *ſhortned* the *Difficulty*, becauſe it ſpared them the Trouble of accounting for their *Conveyance* from *Sea*, which was what had ſo ſeverely exerciſed all the former. Though, in reality, this

this only *heighten'd* and *enhansed* it: and render'd it still more *intricate*; as does in some Measure already, and will hereafter appear more at large, when I shall have publish'd the *preliminary Dissertation*, whereof I have given some Account above. And *this* was the *most received* and *prevalent Opinion* * when I first brought my *Collection* of these *Things* up to *London*.

* *Vid. Ray's Three Discourses* 8°. *Lond.* 1693, *pag.* 127.

There have been, besides these recited, some *other Conjectures* propos'd about the removal of these Bodies to Land; which I choose, rather than trouble the *Reader* with a Detail of them here, to deferr to their proper Place, that I may proceed directly onwards in my *Design*. Now the more effectualy to smooth my *Way*: and that this very great *Diversity* of *Opinions* may not be any longer an Amusement to the *World*, 'twill be very convenient that I look into the *Reasons* and *Pretensions* of each: and shew upon what *Ground* 'tis that I embrace that of the *Deluge*, and set aside *all the rest*.

Why I adhere to them who suppose these *Marine Productions* brought

 out

out by the *Univerſal Deluge*, will be beſt learn'd from the *ſucceeding Part* of this *Eſſay*, which is wholey dedicated to that Purpoſe. And to that I ſhall prefix an *Hiſtorical* Account of the *Labours* of *Fab. Columna*, *Nic. Steno*, *P. Boccone*, *Jac. Grandius*, Mr. *John Ray*, and other *Learned Men*, on *this Subject*; ſhewing what they have *already done* in it, wherein they *failed*, and what remains *ſtill* to be *done*.

Why I *reject* all the other *Conjectures*, falls under our *preſent Conſideration*. And, to make as *ſhort* of the Matter as poſſible, 'tis becauſe they will none of them abide the *Teſt*. Becauſe they have not due Warrant from *Obſervation*, but are clearly repugnant thereunto. In a Word, becauſe the *preſent Circumſtances* of theſe *Marine Bodies* do not ſquare with *thoſe Opinions*, but exhibit *Phænomena* that *thwart* them, and that give *plain Indications* that they could never have been put into the *Condition* we now find them by any ſuch *ſhort* and *partial Agents* as thoſe they propoſe.

Now

Now in regard that the *ſaid Circumſtances* are impartialy related in my *Obſervations*, we need only have recourſe to *them* to put an *End* to this *Buſineſs*. For, as *Mathematicians* ſay of a *ſtreight Line*, that 'tis as well an *Index* of its own *Rectitude*, as of the *Obliquity* of a *crooked* one ; ſo *theſe* may ſerve indifferently to detect the *erroneous Ways*, and to point forth the *true*. And it is from theſe *Obſervations* *: from the *Number*, *Order*, *Variety*, *Situation*, *Depth*, *Diſtance* from the *Sea*, and *other Accidents* of *theſe Bodies* *, that I ſhall ſhew,

That *they* were not brought, from *Sea*, to the *Parts* where they are *now found*, by *Men*, the *ancient Inhabitants* of thoſe *Parts*, as *ſome Authors* have been of *Opinion* ; they preſuming that theſe *Shells* were at firſt only flung out upon the *Surface* of the *Earth* : and that thoſe which we now find *buried* in it, were, in tract of Time, *cover'd*, either by that *Terreſtrial Matter* which falls down along with *Rain*, or by the 1.

* * See a brief Detail of theſe *Obſervations* in the Beginning of Part II.

Earth which is waſh'd from off the *Hills* by *Land Floods*.

2. That *they* were not *carry'd*, together with the *Water* which ſome ſuppoſe to paſs, continualy, from the Bottom of the *Sea* to the Heads of *Springs* and *Rivers*, through certain *ſubterranean Conduits* or *Chanels*, until the Shells were by ſome Glut, Stop, or other Means arreſted in their *Paſſage*, and ſo detained in the *Bowels* of the *Earth*; as *others* have rather inclined to believe.

3. That *they* were not *born forth* of the *Sea*, and laid upon the *Land* by particular *Inundations*; ſuch as were the *Ogygean*, the *Deucalionean*, and others of freſher Date: ſuch as are *thoſe* which uſualy attend *Earthquakes*: or thoſe which are ſometimes occaſioned by very high *Tides*, by impetuous *Winds*, and the like; as *other Writers* have thought.

4. That *they* were not left behind at the *Beginning* of the *World*, when the *Sea* overſpread the *whole Globe*, till its *Retreat* into its aſſigned *Chanel*, and *the Waters were gathered together unto one place*, * the third Day

* Gen. 1. 9.

Day from the Commencement of the *Creation*; which *others* believ'd.

That *they* were not left by the *Seas* being conſtrained to *withdraw* from off certain Tracts of *Land*, which lay till then at the *Bottom* of it, but being *raiſed* to an higher Pitch, ſo as to ſurmount the Level of the *Seas Surface*, they, by that Means, became *Iſlands* and *habitable*; the ſaid Tracts being thus elevated by *Earthquakes*, or the like ſubterraneous Exploſions; in ſuch Manner as *Rhodes*, *Thera*, *Theraſia*, and many other *Iſlands* were ſuppoſed to have been raiſed; which is the *Conjecture* of *others*. 5.

That *they* were not left by the *Sea's changing* its *Place*, receding from the *Parts* it *anciently poſſeſs'd*, and betaking it ſelf to *new Quarters*; *this Change* being occaſion'd by ſome accidental *Emotion* or *Tranſpoſition* of the common *Center* of *Gravity* in the *Terraqueous Globe*; and thereupon the *Fluids* of it, the *Sea*, and the reſt, immediately *ſhifting* likewiſe, as being the more eaſily moveable Parts of the Maſs, and coming to another *Æquilibrium*, 6.

that

that they might thereby the better accommodate themſelves to their *new Center*; as *others*.

7. That *they* were not left upon the *Sea's* being *protruded* forwards, and conſtrained to *fall off* from certain *Coaſts*, which it formerly poſſeſſed, by the *Mud* or *Earth* which is diſcharged into it by *Rivers*; the ſaid *Mud* being repos'd along the *Shores* near the *Oſtia* of thoſe Rivers, and by that Means making continual *Additions* to the *Land*, thereby *excluding* the *Sea*, dayly invading and *gaining* upon it, and preſerving theſe *Shells* as *Trophies* and *Signs* of its *new Acqueſts* and *Encroachments*; which *Others* have imagined; they concluding that the Iſlands *Echinades*, the Lower *Egypt*, *Theſſaly*, and many other *Countries*, were thus rais'd out of the *Mud* brought down by *Achelous*, the *Nile*, *Peneus*, and other *Rivers*.

8. Laſtly. That *they* were not left by the *Sea's* continual *flitting* and *ſhifting* its *Chanel*, this *Progreſſion* being occaſion'd by the *Sea's wearing* and *gaining* upon *one Shore*, and flinging up *Mud*, and, together with it, theſe *Shells*,

Shells, upon the *other*, or oppoſite *Coaſts*, thereby making perpetual *Additions* unto them; which is the *Opinion* of *other Authors*.

Theſe *Propoſitions* (which are no other than ſo many *Conſectaries* drawn from the *Obſervations*) are, we ſee, all *Negative*; as being directed againſt the *Miſtakes* of *ſome* who have formerly engag'd in this *Reſearch*. The *Ways* they have taken to account for the *Conveyance* of theſe *Marine Bodies* to *Land*, are very *many*, as well as *different* from each other. For ſo *eager* and *ſollicitous* hath the *inquiſitive* and *better* Part of *Mankind* been to bring this Matter to a fair *Iſſue* and *Determination*, that no Stone hath been left unturned, no Way, whereby theſe Things could ever poſſibly have been brought forth of the *Sea*, but one or other of them hath pitch'd upon it. So that, by this *Refutation* of all theſe, I might prove my *own* (which is the only one remaining) by *Induction*; but this kind of *Proof* is not needfull, where more *cogent* and *poſitive Arguments* are not wanting.

And

And thus much of *this Part* I get over by the ſole Guidance of my *Senſes*. A *View* of the *preſent State* of theſe *Bodies* alone convinced me ſufficiently that the *Means*, propoſed by *theſe Authors*, were not the *true ones*: that they were both *levell'd wide*, and fell all *ſhort* of the *Mark*. Now tho' this was enough for my preſent Purpoſe; and when I had evinced that, although ſuch *Alterations*, as thoſe which *theſe Gentlemen* ſuppoſe, *Tranſitions*, and *Migrations* of the *Center* of *Gravity*: *Elevations* of new *Iſlands*: whole *Countries* gain'd from the *Sea*: and other like *Changes*, had *actualy* happen'd, yet theſe *Shells* could never poſſibly have been *repoſited* thereby in the Manner we now find them; I ſay, when I had proved *this*, I was not immediately concerned to inquire whether ſuch *Alterations* had realy ever happen'd or not. Yet, partly for a fuller and more effectual *Diſproof* of the recited *Opinions*: and partly becauſe I am more eſpecialy obliged by my *general Deſign* to look into all Pretences of *Changes* in the *Globe* we inhabit, and I ſaw very

very well that ſcarce *any*, of all theſe alledged, had the leaſt Countenance either from the *preſent Face* of the *Earth*, or any credible and authentick *Records* of the *ancient State* of it, I reſolved to purſue this Matter ſomewhat *farther*, and to ſhew that, although there do indeed happen ſome *Alterations* in the *Globe* yet they are very *ſlight* and almoſt *imperceptible* *, and ſuch as tend rather to the *Benefit* and *Conſervation* of the *Earth* and its *Productions*, than to the *Diſorder* and *Deſtruction* both of the *one* and the *other*, as all theſe *ſuppoſititious ones* moſt manifeſtly would do, were there realy any ſuch. But from clear and inconteſtible *Monuments* of *Antiquity*: from *Hiſtory* and *Geography*: and from attentive *Conſideration* of the *preſent State* of thoſe *Countries* where theſe *Changes* were ſuppoſed to have been wrought, I prove that they are *imaginary* and *groundleſs*, and that *ſuch* in earneſt never happened; but that the *Bounds* of *Sea* and *Land* have been more *fix'd* and *permanent*: and, in ſhort, that the *Terraqueous Globe* is to

* *Confer* Part 5. *Conſ.* 1. &c.

to *this Day* nearly in the ſame *Condition* that the *Univerſal Deluge* left it; being alſo like to *continue* ſo till the Time of its final *Ruin* and *Diſſolution*, preſerved to the *ſame End* for which 'twas firſt *formed*, and by the *ſame Power* which hath *ſecur'd* it hitherto. But, with reſpect to my *preſent Deſign*, I more particularly make out,

That although *Rain-water* be indeed (as theſe *Writers* ſuppoſe) very plentifully *ſaturated* with *Terreſtrial Matter*, and (as I ſhall make appear) that *peculiar Matter* out of which the *Bodies* of *Vegetables*, and conſequently of *Animals*, are *formed*, *nouriſh'd*, and *augmented*, *Water* being the common *Vehicle* and *Diſtributer* of it to the Parts of thoſe *Bodies*, and *all Water* (eſpecialy that of *Rain*) being, more or leſs, *ſtored* with this, it being *light* in compariſon of the common *mineral earthy Matter*, and therefore eaſyly aſſumed into *Water*, and *moved* along with it; yet that this *Matter* being all originaly derived from the *Surface* of the *Earth*, either by the *Vapour* that continualy *iſſues* out, and

and *ascends* from all Parts of it †, or *wash'd* off by *Land-floods*, and conveyed into *Rivers* and the *Sea*, and thence *elevated* up, together with the *Vapour*, which, as the former, constitutes the *Rain* that falls; I say, it being thus *originaly* all *rais'd* from the *Earth*, when *restor'd back* again thereunto, 'tis but where it was *before*, and does not *enlarge* the *Dimensions* of the *Globe*, or augment the *Surface* of the *Earth*, and only lye idley and unserviceably there, but *Part* of it is *introduced* into the *Plants* which grow thereon, for their *Nutrition* and *Increment*, and the *rest*, which is superfluous, either *remounts* again, with the ascending *Vapour*, as before, or is wash'd down into *Rivers*, and transmitted into the *Sea*, and does not make any sensible *Addition* to the *Earth*, whatever *Some* may have believ'd.

† *Part* 3. *Sect.* 1. *Cons.* 8.

That the *Terrestrial Matter*, which is thus carry'd by *Rivers* down into the *Sea*, is *sustained* therein, partly by the greater *Crassitude* and *Gravity* of the *Sea-water*: and partly by its constant *Agitation*, occasion'd by the *Tides*, and by its other *Motions*; and

and is not permitted to *ſink* to the *Bottom.* Or, if any of it do, 'tis *rais'd* up again by the next *Storm :* and, being *ſupported* in the Maſs of *Water*, together with the reſt, 'tis by Degrees *exhaled*, mounted up with the *Rain* that *riſes* thence, and returned back again to the *Earth* in fruitul *Showers.* By this perpetual *Circulation* a great many Things in the *Syſtem* of *Nature* are tranſacted : and two main *Intentions* of *Providence* conſtantly promoted; the *one* a *Diſpenſation* of *Water* promiſcuouſly and indifferently to all *Parts* of the *Earth*; this being the immediate *Agent* that both *bears* the *conſtituent Matter* to all formed *Bodies*, and, when brought to them, *inſinuates* it in, and *diſtributes* it unto the ſeveral *Parts* of thoſe *Bodies*, for their *Preſervation* and *Growth :* the *other*, the keeping a juſt *Æquilibrium*, if I may ſo ſay, betwixt the *Sea* and *Land*; the *Water*, that was raiſed out of the *Sea*, for a *Vehicle* to this *Matter*, being by this Means *refunded back* again into it, and the *Matter* it ſelf reſtored to its original *Fund* and *Promptuary*, the

the *Earth*; whereby each is *restrained*, and kept to due *Bounds*; so that the *Sea* may not *encroach* upon the *Earth*, nor the *Earth* gain Ground of the *Sea*. That there never were any *Islands*, or other like Parcels of *Land*, amassed or heap'd up: nor any considerable *Inlargement*, or Addition of Earth made to the *Continent*, by the *Mud* that is carryed down into the Sea by Rivers. That although the *Ancients* were almost unanimously of Opinion that those Parts, where *Egypt* now is, were formerly *Sea*: and that a very large Portion of *that Country* was *recent*, and form'd out of the *Mud* discharg'd into the neighbouring Sea by the *Nile*, yet this Tract of Land had no such *Rise*, but is as *old*, and of as long a Standing, as *any* upon all the whole Continent of *Africa*: and hath been in much the *same natural Condition*, that it is at *this Day*, ever since the Time of the *Deluge*; its *Shores* being neither *advanced* one jot further into the *Sea* for this three or four thousand Years, nor its *Surface* raised by additional *Mud* deposed upon it by the yearly *Inundations* of the *Nile*. That neither the *Palus Mæotis*, nor the *Euxine*, nor any other *Seas*, fill up, or by Degrees grow *ſhallower*. That *Salmydeſſus*, *Themiſcyra*, *Sidene*, and the adjacent Countries, upon the Coaſts of the *Euxine Sea*, were not formed out of the *Mud* brought down by the *Iſter*, *Thermodon*, *Iris*, and the other Rivers which diſcharge themſelves into *that Sea*. That *Theſſaly* was not raiſed out of the *Mud* born down by the River *Peneus*: the Iſlands *Echinades*, or *Curzolari*, out of that brought by the River *Achelous*: *Cilicia*, by the River *Pyramus*: *Myſia*, *Lydia*, *Ionia*, and other Countryes of *Anatolia*, by the *Caicus*, *Hermus*, *Cayſter*, and the other *Rivers* which paſs through them. To be ſhort, That no *Iſland* or *Country* in the *whole World* was ever raiſed by *this Means*, notwithſtanding that very many *Authors*, and ſome of conſiderable *Note*, have believed that all the abovementioned *Countryes* were ſo raiſed. Nay, to ſo ſtrange a Height of Extravagance do ſome, otherwiſe Learned and Curious Perſons

ſons run, when they indulge *Fancy* too far, and rely wholey upon *Probabilities* and *Conjectures*, there is hardly any one ſingle *Iſland*, or *Country*, all round the *Globe*, that one *Writer* or other hath not thought to have been formed after this Manner, or, at leaſt, ſome very *large Part* of it.

That there is no *authentick* Inſtance of any conſiderable *Tract* of *Land*, conſiſting, as *uſual* *, of *Strata* †, that was *rais'd* up from the *Bottom* of the *Sea* by an *Earthquake*, ſo as to become an *Iſland*, and be render'd *habitable*. That *Rhodus*, *Thera*, *Theraſia*, and ſeveral other *Iſlands*, which were ſuppoſed by the *Antients*, and, upon their *Authority*, by *later Authors*, to have been thus *raiſed*, had realy no ſuch *Original*: but have *ſtood out* above *Water* as long as the reſt of their Fellow *Iſlands*, and ſtand now juſt as the *Univerſal Deluge* left them.

E 2 That

* *Conf. Part* 2. *Conſ.* 3. *infra.*

† And not a confuſed Heap of *Cinders* thrown forth of ſome *Vulcano*, like that caſt up, a few Years ago, in the Bay of *Santorini*; of which ſee my *Anſwer* to the learned *Camerarius*, *Part*. 2. *Sect*. 8. Of the *Origin* of the *Monte di Cinere* ſomething is noted towards the End of the 2d Part of *this Eſſay*.

That as to that *Affection* of *Bodyes* which is called their *Gravity*, it clearly ſurpaſſes all the *Powers* of meer *Nature*, and all the *Mechaniſm* of *Matter*. That as any one *Body*, or Part of *Matter*, cannot be the *Cauſe* of its *own Gravity*: ſo no more can it ever poſſibly be the *Cauſe* of the *Gravity* of *another Body*, or Part of *Matter*. That neither the *Earth*'s diurnal *Revolution* upon its *Axis*: nor any *magnetick Effluvia* of the *Earth*: nor the *Air*, or *Atmoſphere* which environs the *Earth*: nor the *Æther*, or *Materia ſubtilis* of the *Carteſians*, in what Manner ſoever *moved* or *agitated*: (all which have been propoſed by ſeveral *Learned Men* as the *Cauſes* of *Gravity*) nor any other *Fluid* or *Matter* whatever, can of it ſelf produce ſuch an *Effect* as is that of the *Gravity* of *Bodyes*. That it does not proceed from the *Efficiency* of any ſuch *contingent* and *unſtable Agents*, but ſtands on a *Baſis* more *firm* and *ſtedfaſt*; being entirely owing to the direct *Concourſe* of the *Power* of the *Author of Nature*, immediately in his *Hand*, and the *main Engine*

Engine whereby this ſtupendous *Fabrick* of the *Univerſe* is *managed* and *ſupported:* the *prime Hinge* whereon the whole *Frame* of *Nature* moves: and is *principaly concerned*, if not the *ſole Efficient* in the moſt remarkable *Phœnomena* of the *Natural World*; which, ſhould *Gravity* once *ceaſe*, or be *withdrawn*, would inſtantly *ſhiver* into Millions of *Atoms*, and fall into the greateſt *Diſorder* and *Confuſion*. That the common *Centre* of *Gravity* in the *Terraqueous Globe* is *ſteady*, *immoveable*, and not liable to any accidental *Tranſpoſition:* nor hath it ever *ſhifted* or *changed* its *Station*. And that there is no *Declination* of *Latitude:* nor *Variation* of the *Elevation* of the *Pole*; notwithſtanding what ſome *learned Men* have aſſerted.

What concerns the raiſing of *new Mountains: Deterrations*, or the *Devolution* of *Earth* down upon the *Valleys*, from the *Hills* and higher Grounds: and *Iſlands* torn off from the main *Continent* by *Earthquakes*, or by the furious and impetuous *Inſults* of the *Sea*; theſe, I ſay, will fall more properly under our

Conſideration on another Occaſion*. And for the *Mutations* of leſſer moment, which ſome have fancied to have happen'd within this *Intervall*, I mean, for the laſt four thouſand Years ſince the Deluge, I chuſe rather to paſs them over at preſent, than to crowd and encumber this *ſhort Tract* with the Account of them.

* *Part* 2. and *Part* 5.

I muſt needs freely own, that when I firſt directed my Thoughts *this Way*, 'twas Matter of real Admiration to me, to find that a Belief of *ſo many* and ſo *great Alterations* in the *Earth*, had gained ſo large Footing, and made good its Ground ſo *many Ages* in the *World*; there being not the leaſt *Signs* nor *Footſteps* of any ſuch Thing upon the Face of the *whole Earth*: no tolerable *Foundation* for ſuch a *Belief* either in *Nature*, or *Hiſtory*. But I ſoon ſaw very well, that the *Moderns* generaly entertain'd it meerly upon the Credit and Tradition of the *Ancients*, and *that* without due Examination, or Enquiry into the *Truth* and *Probability* of it. And 'twas not long e'er I diſcovered what it it was that did ſo generaly *miſlead* the *Ancients* into theſe *Miſtakes.* But of that more by and by.

Thoſe *ancient* Pagan *Writers* were indeed very much *excuſable* as to this Matter. *Philoſophy* was then again in its *Infancy*; there remaining but few Marks of the old *Tradition*, and thoſe much obliterated and defaced by *Time.* So that *they* had only dark and faint *Idea's*, narrow and ſcanty Conceptions of *Providence:* and were ignorant of its *Intentions*, and of the *Methods* of its *Conduct* in the Government and Preſervation of the *Natural World.* They wanted a longer *Experience* of *theſe Things:* a larger Stock of *Obſervations*, and *Records* of the *State* of the *Earth* before their *Times*; having, as Things then ſtood, nothing to aſſiſt them in their *Enquiries* beſides their own *Gueſſes* and *Fancy.* For their *Progenitors*, and thoſe who lived in the *earlyer Ages*, were almoſt entirely taken up with *Buſineſs* of another Kind. That fatal Calamity, the *Deluge*, had wrought ſuch a *Change* *, that they beheld every where a *new Face* of

* *Confer* *Part* 2.

 Things:

Things: and the *Earth* did not *then* teem forth its *Encreaſe*, as *formerly*, of its *own Accord*, but required *Culture*, and the *Aſſiſtance* of their Hands, much more than before it did. The Proviſion of Bread for *Food*: *Clothing* to ward off the *Injury* and *Inclemency* of the *Air*: and orher like *Employs* for the *Comfort* and *Support* of *Life*, being of indiſpenſible Neceſſity, were to be firſt look'd after. And theſe *Employs*, being then for the moſt part *new* to them, and ſuch as they were *unſkill'd* in, were alone enough to take up the greateſt part of their Time. The *Methods* they uſed of *Agriculture*, and other *Arts* of like Importance, were ſo aukward and tedious, as to afford them little *Leiſure* for *Works* of the *Brain*, for *Hiſtory*, or Contemplations of that Nature. And till better *Experience* had led their *Poſterity* to the *Improvements* of *Arts*: till the *Plow*, and other uſeful *Inſtruments*, were found out: and they had learn'd more compendious and expeditious Ways of diſpatching thoſe *Affairs*, whereby they *ſhortned* their *Labours*,

and

and ſo *gained Time*, there was no *Shew* of *Learning*, or Matters of Speculation among them; and we hear little or nothing of *Writing*; nay 'twas a very conſiderable Time before *Letters* themſelves were found out. I know very well, there are ſome who talk of *Letters* before the *Deluge*; but that is a Matter of meer *Conjecture*, and I think nothing can be peremptorily determined either the one Way or the other; though I ſhall ſhew, that 'tis highly *probable* they had *none*. Be *that* how it will, I ſhall plainly make out, that the *Ages* which next ſucceeded the Deluge *had none*. Indeed they knew nothing at all of them: and the *firſt Writing* they uſed was only the ſimple *Pictures*, and *Gravings* of the *Things* they would *repreſent*, Beaſts, Birds and the like; which Way of Expreſſion was afterwards call'd *Hieroglyphick*. But *this* fell into *diſuſe*, when *Letters* were afterwards *diſcover'd*; they being, in all Reſpects, a far more *excellent* and uſefull *Invention*. We ſee therefore that there were ſeveral *Reaſons* why thoſe *early Ages*

ges could not tranſmitt Accounts of the *State* of the *Earth*, and of theſe *Marine Bodies*, in *their Times*, down to the *ſucceeding Generations.* So that *theſe* having little more to truſt to than their *own Imagination*, and no ſurer a *Guide*, in their *Reaſonings* about *theſe Things*, than *bare Conjecture*, 'twas no Wonder that they fell into groſs and palpable *Miſtakes* concerning them.

Nor much more Wonder is it that *Epicurus*, who could ever eſpouſe a *Notion* ſo enormouſly abſurd, and groundleſs, as that the World was framed by *Chance:* that this vaſt, regular, and moſt ſtupendous *Pile* was owing to no higher a *Principle* than a *fortuitous Congreſs of Atoms:* and that either there was no *God* at all, or, which is much the ſame Thing, that he was an impotent and lazy *Being*, and wholey without *Concern* for the *Affairs* of this *lower World*; I ſay, 'tis in no wiſe ſtrange that ſuch a One ſhould believe, as he did, that *Things* were blindly ſhuffled and hurled about in the *World:* that the *Elements* were at conſtant *Strife* and *War* with each

each other: that, in ſome Places, the *Sea* invaded the *Land:* in others, the *Land* got Ground of the *Sea:* that *all Nature* was in an *Hurry* and *Tumult:* and that as the *World* was firſt *made*, ſo ſhould it be again *diſſolved* and *deſtroyed* by *Chance:* that it had already made large *Advances* that Way, being *infirm* and *worn* with *Age*, ſhatter'd and *crazy*, and would in Time dwindle and relapſe again into its ſuppos'd original *Chaos.*

Did *Gravity*, the Inclination of *Bodyes* towards a *Center*, to which Inclination they owe their reſpective *Order*, and *Site* in regard of each other, very many of their *Motions* and *Actions*, and in a great meaſure, their preſent *Conſtitution*; did *this*, I ſay, happen from ſo contingent, precarious, and inconſtant *Cauſes* as many have believ'd: or did it ſtand upon ſo ticklish and tottering a *Foundation* as ſome Mens Fancy hath placed it, 'twould be no Wonder ſhould it frequently *vary:* its *Center* ſwerve and *ſhift*, upon every turn: and that there ſhould enſue thereupon, not only ſuch *Motions*, and *Alterations* of the *Bounds*

Bounds of the *Sea* as they imagine; but likewiſe many other, and not leſs pernicious *Perturbations* of the Courſe of even *univerſal Nature.*

Or was the *Univerſe* left to its *own Conduct* and *Management:* the whole Maſs of created *Matter* to its *proper Diſpoſition* and *Tendency:* were there no Reſtraint of *Bounds* to the *Earth*, nor *Curb* to the *Ocean:* was there not *One* who had *ſet Bars and Doors to it, and ſaid hitherto ſhalt thou come, but no farther, and here ſhall thy proud Waves be ſtaid* *; then indeed might we well expect ſuch *Viciſſitudes* and *Confuſions* of Things: ſuch Juſtlings and Claſhings in *Nature:* ſuch Depredations, and *Changes* of *Sea* and *Land.*

* *Job* xxxviii. 10, 11.

But if the ſame mighty *Power*, which in the *Beginning* produced this vaſt *Syſtem* of *Bodyes* out of *Nothing*, and diſpoſed and ranged them into the moſt excellent and beautiful *Order* we now behold: which at firſt framed an *Earth* of a *Conſtitution* ſuitable to the *innocent State* of its *primitive Inhabitants:* and afterwards, when *Man* had *degenerated*, and quitted that *Innocence*, alter'd alter'd that *Constitution* of the *Earth*, by means of the *Deluge* †, and reduced it to the *Condition* 'tis now in, thereby adapting it more nearly to the *present Exigencies* of Things, to the *laps'd* and *frail State* of *humane Nature* : If that same *Power* be *yet* at the *Helm* : if it *preside* in the *Government* of the *Natural World* : and hath still the same peculiar *Care* of *Mankind*, and, for *their Sake*, of the *Earth*, as heretofore, (all which shall be evidently made out,) then may we very reasonably conclude 'twill also *continue* to *preserve* this *Earth*, to be a convenient *Habitation* for the *future Races* of *Mankind*, and to furnish forth all Things necessary for their *Use*, *Animals*, *Vegetables*, and *Minerals*, as long as *Mankind* it self shall endure ; that is, till the *Design* and *Reason* of its *Preservation* shall cease. And till then, so *steady* are the *Purposes* of *Almighty Wisdom* : so firm, establish'd, and constant the *Laws*, whereby it *supports* and *rules* the *Universe* ; the *Earth*, *Sea*, and all *natural Things* will continue in the *State* wherein they now are, without the least *Senescence*

† *Part* 2. *Consf.* 2. *&c.*

Seneſcence or *Decay:* without *Jarring*, *Diſorder*, or *Invaſion* of one another: without Inverſión or Variation of the ordinary Periods, Revolutions, and Succeſſions of Things: and we have the higheſt Security imaginable, that *While the Earth remaineth*, *Seed-time and Harveſt*, *and Cold and Heat*, *and Summer and Winter*, *and Day and Night*, *ſhall not ceaſe**.

* Gen. viii. 22.

And whatever may be urged in Behalf of the *Ancients*, I cannot well ſee, I confeſs, what can be ſaid for the *later Authors*, who have embrac'd the *ſame Tenets*, more than that theſe Learned Men took up thoſe Tenets *on Truſt*; their over-great Deference to the Dictates of *Antiquity* betraying them into a Perſuaſion of ſuch *Changes* in the *Earth*. I have given my *Reaſons* above why I cannot think the Antients competent *Judges* in this *Caſe*. We have, at this time of Day, better and more certain *Means* of *Information* than *they* had: and therefore it were to have been wiſh'd that theſe Gentlemen had not thus obſequiouſly followed *them*, but gone another Way to work. It would certainly have

have been much better, had they taken the Pains to have look'd a little into *Matter of Fact:* had they consulted *History* and *Geography*, in order duly to acquaint themselves with the *past* and *present State* of the *terraqueous Globe:* and not to have pass'd *Sentence* till they had first compared the most *antient Descriptions* of *Countries* with the *Countries themselves* as *now* they stand. Nay, had they but read and attended to the *Accounts* which the very *Authors*, from whom they borrow these *Opinions*, have left us, they might have discover'd, even from *them*, the *Errors* and *Oversights* of their *Authors:* and have learned, that the *Face* of *Sea* and *Land* is the very *same* at *this Day* that it was when those *Accounts* were compil'd: and that the *Globe* hath not sustain'd any considerable *Alterations*, either in the *Whole*, or any of its *Parts*, in all this time.

Those who can content themselves with a *Superficial View* of *Things:* who are satisfy'd with contemplating them in *Gross:* and can acquiesce in a general, and less nice *Examination*

mination of them: whoſe *Thoughts* arc *narrow* and *bounded:* and their *Proſpects* of *Nature* ſcanty, and by Piecemeal, muſt needs make very *ſhort* and defective *Judgements*, and, oftentimes very *erroneous*, and wide of Truth. Some fancyful Men have expected nothing but *Confuſion* and *Ruin* from *thoſe very Means* whereby both *that* and *this* is moſt effectualy *prevented* and *avoided*. *One* imagines that the *terreſtrial Matter*, which is ſhowred down along with *Rain*, enlarges the *Bulk* of the *Earth*, and that it will in Time *bury*, and lay all Things *under Ground*. *Another*, on the contrary, fancyes that the *Earth* will e'er long all be *waſh'd away* by *Rains*, and born down into the *Sea* by *Rivers:* and, its *Chanel* being thereby quite *filled* up, the *Waters* of the *Ocean* will be turned forth to *overwhelm* the *dry Land*. Whereas by this *Diſtribution* of *Matter*, continual *Proviſion* is every where made for the *Supply* of *Bodies:* the juſt *State of Sea* and *Land* preſerved, and the *Bounds* of each ſecur'd; quite contrary to the prepoſterous *Reaſonings* of thoſe Men, who

who expected ſo different a *Reſult* of theſe Things. And ſhould this *Circulation*, from which they dreaded thoſe diſmal *Conſequences*, once ceaſe, the *Formation* of *Bodyes* would be immediately at an *End:* and *Nature* at a perfect *Stand.* But I am aware that I tranſgreſs: and that this is a *Prolixity* not allowable in a *Treatiſe* of this Nature; wherefore I ſhall *conclude*, after I have performed my *Promiſe* of diſcovering what it was which led the *ancient Hiſtorians*, *Geographers*, and Others ſo generaly into a Belief of theſe frequent *Changes* betwixt *Sea* and *Land:* and 'twas *this.*

They obſerved, almoſt wherever they caſt their Eyes, vaſt *Multitudes* of *Sea-ſhells*, at *Land*, in their *Fields*, and even at very great *Diſtance* from any *Sea.* This, *Erataſthenes*, *Herodotus*, *Xanthus Lydus*, *Strabo*, *Pauſanias*, *Pomponius Mela*, *Theophraſtus*, *Strato* the Philoſopher, *Plutarch*, and others of them aſſure us. They found them upon the *Hills*, as well as in the *Valleys*, and *Plains:* they obſerved that they were immerſed in the *Maſs* of the *Stone* of their

Rocks, *Quarries*, and *Mines*, in the ſame Manner as they are at this Day found in all known *Parts* of the *World*.

Nay, in thoſe *Elder Times*, and which were ſo much *nearer* to the *Deluge* than ours are, they found theſe *Marine Bodies* more *frequently*, and in much *greater Plenty*, than we *now* do: and moſt, if not all of them, *freſh*, *intire*, and *firm*. The whole *cruſtaceous Kind*, and the *lighter ones* of the *teſtaceous*, which together would be a *vaſt Number*, ſubſiding laſt, fell upon the *Surface* of the *Earth* *; whilſt the *heavier*, which ſettled down before, were entombed in the *Bowels* of it. *Thoſe* therefore muſt *then* lye every where ſtrew'd upon the Ground. Whereas *now* very few, if any, of them appear †; the Shells which we find at *preſent* upon the Face of the Earth being principaly of the *heavier Sorts*, which were at *firſt* lodged *within it*, and *ſince* diſcloſed and turned out; by what Means we ſhall ſee hereafter ||. And indeed, 'tis not conceivable how the Generality of them could *endure* ſo many Hun-

* *Pag.* 30, & *ſeqq.* and *Pt.* 2. *Conſ.* 3.

† *Conf.* *p.* 34. and *Part* 2. *Conſ.* 3.

|| *Part.* 5. *Conſ.* 4.

Hundreds of Years as have ſince paſt: how they could lye ſo long expoſed to the *Air*, *Weather*, and other *Injuries*, without vaſt Numbers of them, and eſpecialy the finer and tenderer Species, being, long e'er this, *periſh'd* and *rotten:* ſome of them quite *deſtroy'd* and *vaniſh'd*, and the reſt ſo *damag'd*, many of them, and *alter'd* by *Time*, as not to appear the Things they then were, and ſo create a Doubt amongſt ſome of *us* whether they are realy *Shells* or not.

This was a *Scruple* that never enter'd into *their Heads*. The *Shells*, being then fair, *ſound*, and free from *Decay*, were ſo exactly *like* thoſe they ſaw lying upon their *Shores*, that they never made any Queſtion but that they were the *Exuviæ* of *Shell-fiſh:* and that they once belonged all to the *Sea*. But the *Difficulty* was how they came *thither:* and by what Means they could ever arrive to *Places* oftentimes ſo *remote* from the *Ocean*.

The *Ages* that went *before* knew well enough how theſe *Marine Bodies* were brought thither. But ſuch were the *Anxieties* and *Diſtreſſes* of the then again *infant World:* ſo inceſſant their *Employs* about Proviſion for *Food*, *Rayment*, and the like, that (even after *Letters* were diſcover'd) there was little *Leiſure* to committ any thing to *Writing:* and, for want *thereof*, the *Memory* of this extraordinary *Accident* was in great Meaſure *worn out* and *loſt*. 'Tis true there was a general and loud *Rumour* amongſt them of a mighty *Deluge* of Water that had drown'd all *Mankind*, except only a very few *Perſons*. But there had alſo happen'd very terrible *Inundations* of later *Date*, and which were *nearer* to the *Times* when theſe *Authors* lived. Such was that which overflow'd *Attica* in the Days of *Ogyges:* and that which drowned *Theſſaly* in *Deucalion*'s Time. *Theſe* made cruel *Havock* and *Devaſtation* amongſt them: their own native Country, *Greece*, was the *Theatre* whereon theſe *Tragedies* were acted, and their *Progenitors* had ſeen and felt their *Fury*. And *theſe* happening nearer Home, and their *Effects* being freſh, and in all Mens Mouths, they made

ſo

ſo ſenſible and laſting *Impreſſions* upon their *Minds*, that the old great *Deluge* was eclipſed by that Means, its *Tradition* mightily obſcured, and the *Circumſtances* of it ſo interwoven and confounded with thoſe of theſe *late Deluges*, that 'twas e'en dwindled into *Nothing*, and almoſt bury'd in the *Relations* of *thoſe Inundations.*

In *their Inquiries* therefore into this *Matter*, ſcarcely a Man of them thought, or ſo much as dream'd of the *Univerſal Deluge.* They concluded indeed unanimouſly that the *Sea* had been *there*, wherever they met with any of theſe *Shells*: and that *it* had left them behind. And ſo far they were in the *right*: *this* was an *Inference* rational and natural enough. But when they began to reaſon about the *Means* how the *Sea* got thither, and away back again, *there* they were perfectly in the Dark. And, both *Tradition* and *Philoſophy* failing them, they had recourſe to *Shifts*, and to the beſt *Conjectures* they could think of; concluding that *it* was either forced forth, as in *particular Inundations*, ſuch as thoſe lately mention'd: or that *thoſe Parts*, where they found the *Shells*, had been *formerly* in the *Poſſeſſion* of the *Sea*, and the Place of its natural *Reſidence*, which it had ſince quitted and *deſerted*.

Upon *this* they began to ſeek out by what *Means*, moſt probably, the *Sea* might have been *diſpoſſeſs'd* of thoſe *Parts*, and conſtrained to move into *other Quarters*. And, if 'twas an *Iſland* where they found the *Shells*, they ſtraitways concluded that the *whole Iſland* lay *originaly* at the *Bottom* of the *Sea:* and that 'twas either hoiſted up by ſome *Vapour* from beneath: or that the *Water* of the *Sea*, which *formerly* cover'd it, was in Time *exhaled*, and *dryed* up by the *Sun*, the *Land* thereby laid *bare*, and theſe *Shells* brought to *Light*. But if 'twas in any Part of the *Continent* where they found the *Shells*, they concluded that the *Sea* had been *extruded* and *driven off* by the *Mud* that was continualy brought down by the *Rivers* of thoſe Parts.

That I may not be over-tedious here, I will only add, that I ſhall clearly

clearly ſhew, from plain *Paſſages* of their own *Writings*, yet extant, that 'twas meerly the finding theſe *Sea-Shells* at *Land* that occaſioned this Stir, and raiſed all this Duſt amongſt the *Ancients*: and upon which principaly they grounded their *Belief* of the *Viciſſitudes* and *Changes* of *Sea* and *Land*, wherewith their *Writings* are ſo filled. But how little Reaſon they had for it: and how far *thoſe* have been over-ſeen who have followed them herein, hath been intimated already, and will appear farther from the *following Part* of this *Eſſay*, to the Account of which I now haſten.

PART II.

Concerning the Univerſal Deluge. *That theſe* Marine Bodyes *were* then *left at* Land. *The* Effects *it had upon the* Earth.

THE *Conſectaryes* of the *former Part* of this *Diſcourſe* are all *negative*; *that* being only introductory, and ſerveing but to free the Way to this *ſecond Part*: to reſcue the Enquiry from the *Perplexityes* that ſome *Undertakers* have *incumber'd* it withall; and to ſet aſide the *falſe Lights* they uſed in Queſt of the *Agent* which tranſpoſed theſe *Sea-ſhells* to *Land*.

Now

Now the only *ſure Lights* we have to conduct us, in the *Aſcertaining* this Affair, are Hiſtory of Fact, and *Obſervations*. So that I ſhall give here ſome Intimation of the *Chief* of *thoſe* that ſerve to clear up *this Subject*, and bring the Thing in Queſtion to a fair *Deciſion*. Theſe are, That the *Earth*, all round the *Globe*, appears, wherever it is laid open, to be *wholey compos'd of Strata*, lying on each other, in Form of ſo many *Sediments*, fall'n down, ſucceſſively, from *Water*. That, accordingly, thoſe *Strata* that ly *deepeſt*, are ordinarily the *thickeſt*: and thoſe that ly *above*, gradualy *thinner*, quite up to the Surface. That there are *Sea-ſhells*, and *Teeth* and *Bones of Fiſhes*, found *repoſited* in theſe ſeveral *Strata*; not only in the more lax, *Chalk*, *Clay*, and *Marle*, but even in the moſt ſolid, *Stone*, and the reſt. That theſe *marine Bodyes* are *incorporated* with the *Sand* that conſtitutes the *Stone* of theſe *Strata*, in ſuch ſort as together to *compoſe one common Maſs*. That on Breaking up this Maſs, ſo as to part the *Shell* from the *Stone*, *this* is ever obſerv'd

to

to have receiv'd an *Impreſſion*, of the *exteriour Surface* of the *Shell*, ſo exact as to ſhew it had been *contiguous* and *apply'd* to *all Parts* of the *Shell*; which the *Stone* could not have been capable of, had it not been then in a *State of Solution*, the Matter whereof it conſiſts *looſe*, and ſucceptible of *Impreſſion*. That, upon Breaking the *Shells*, and examining the *Inſides* of them, they are found to contain in them *Stone*, commonly of the ſame Kind with that without, which the *Stratum* is made up of: and *apply'd* as exactly to the *Inſides* of the *Shells*; ſo as to have taken the *Impreſſion* and all the *Lineaments* of them, after the Manner of Matter *caſt*, ſoft, or melted, in a *Mould*. That the *Shells* are, as frequently, *immers'd* in the *Subſtance* of the *Mineral*, and *Metallic Nodules*, even the moſt firm and ſolid, *Flint*, *Spar*, *Pyritæ*, and the reſt; the Matter of theſe *Nodules* exhibiting the *Lineaments* and *Impreſſions* of both the *Outſides* and *Inſides* of the *Shells*, as truely as the *Stoney Matter* of the *Strata* does. That theſe *Marine Productions* are thus *repoſited* as well in in the *loweſt Strata*, as in the *uppermoſt* : at the *Bottoms* of the *deepeſt Mines*, as to the very *Tops* of the *higheſt Mountains*. That they are obſerv'd in ſome Places in ſuch *Multitudes* as, in Bulk, and Quantity, to *equal* if not *exceed* the *Sand*, or other *terreſtrial Matter* of the *Strata*. That there are ordinarily dig'd up, amongſt the reſt, *Shells* that are of *forreign Origin* and *Extract*; being *not the Product* of the *Neighbouring Seas*, but of *Seas* much *remote* and at great *Diſtance*. Thus we here *in England* diſcover, frequently at great *Depths*, *Shells* of Fiſh, very numerous, and of different Kinds, that appear now liveing only on the Coaſts of *Peru*, and other Parts of *America*. That there are likewiſe *diſcover'd*, commonly, at Land, and in the Bowels of the Earth, *Shells* that are not at this Day found *liveing* on any *Coaſts*; being doubtleſs ſuch as naturaly reſide and inhabit only in the *deepeſt* and moſt remote *Receſſes* of the *Main Ocean*, without ever now approaching near any Shore, or being conſequently ever ſeen*. That, in *all*

* *Confer* Pag. 27, & 28. *ſupra.*

all Parts of the *Earth*, as well in *Aſia*, *Africa*, and *America*, as in *Europe*, as well in *Countryes* the moſt *Diſtant* from any *Seas*, as thoſe that by *nearer* to them, the *Strata* are compil'd, and the *Marine Bodyes* diſpos'd in them, every where after the very *ſame Method:* and ſo as apparently to ſhew *Things* were reduced into this *Method*, in *all Countryes*, at the *ſame Time*, and by the *ſame Means*. That there are alſo lodg'd in the *Strata*, *Bones*, *Teeth*, and other Parts, of *Quadrupedes*, or *Land-Animals*, and oftentimes of ſuch as are *not Natives* of the *Country* in which they are thus found. Particularly here in *England* we dig up the *Tuſks*, and the *Grinder-Teeth*, the *Bones*, yea whole *Skeletons*, of very great *Elephants:* and likewiſe incredibly large *Horns* of the *Mooſe Deer*, a Creature not known to be *now liveing* in any other Country excepting *America*. That there are, beſides, repoſited in *Stone*, and even in the firmeſt and hardeſt *Strata*, *Leaves* of various Kinds of *Vegetables:* and ſometimes whole *Trees:* as alſo ſuch *Fruits* as are dureable, firm,

firm, and capable of being preſerv'd, *e. gr. Nuts*, *Pine-Cones*, and the like. That, amongſt the reſt, there are diſcover'd, under ground, *Trees*, *Leaves*, and *Fruits*, of *Vegetables*, in *Countryes* where ſuch do not now ſpontaneouſly *grow*. Nay that there are dig'd up *Trees*, in great *Numbers*, and many of them very *large*, in ſome *Northern Iſlands*, in which there are at this Day *growing* no *Trees* at all : and where, by reaſon of the great *Bleakneſs* and *Cold* of thoſe *Countryies*, 'tis probable none ever did or could grow. That, of all the various *Leaves*, which I have yet ſeen, thus lodg'd in *Stone*, I have obſerv'd none in any other State, nor Fruits further advanc'd in *Growth*, and towards *Maturity*, than they are wont to be at the latter End of the *Spring Seaſon* *. That the ſquamoſe Covers of the *Germina* or Buds, and the Shiv's or Chaff of the *Juli* of *Trees* and *Shrubs*, that fall off in the Spring,

* *When*, according to the *Moſaic Relation*, the *Water* of the *Deluge* came forth, and put a *Stop* to the *Growth* of both *Animals* and *Vegetables*. *Confer Part* 3. *Sect.* 2. *Conſ.* 5. and *Part* 6. towards the End.

Spring, and are found in ſo vaſt Quantityes in many Peat Marſhes, apparently point forth the *ſame Seaſon.* As do likewiſe the immenſe Sholes of the *Ova* of *Fiſhes*, ſo frequent in the upper *Strata* of *Stone.* That the *Shells* of the *Young* of *Fiſh* of the *current Year*, wherever dig'd up, are of the *Size* and *Bigneſs* they are uſed to arrive to at *that Seaſon.* That of all the many *Flyes*, and *Inſects*, that I have yet ſeen inclos'd in *Amber* *, I have never obſerv'd any that were not of the *vernal Tribes* and *Kinds*.

Theſe are the main *Obſervations* whereon I *found* what I *ſhall offer.* Nor is any Thing further needful than a right *Attention* to theſe, to diſcern clearly that the *Methods*, ſet forth in the *precedent Part*, fall far ſhort of *ſolveing* the *Phænomena* here recounted: and that *Obſervation* of the *preſent State* of the *Earth*, and of the *Site* and *Condition* of the *marine Bodyes* in it, is, alone, ſufficient to demonſtrate that they could not

* A Sort of Bituminous Nodule, form'd, with the reſt, during the Deluge, *Vid. Part* IV. *infra*, near the End.

not poſſibly be repoſed, in *that Manner*, by *particular Inundations:* by the *Sea's receding* and *ſhifting* from Place to Place: nor by any of the *other Means* there pretended.

I paſs therefore next on to ſearch out the *true Means:* to diſcover the *Agent* that did actualy bring them forth, and diſpoſe them into the *Method* and *Order* wherein we now find them. To which Purpoſe Nothing is requite, more, than to have Recourſe to ſtill the ſame *Obſervations.* For, by their ſole Aſſiſtance, this Matter may be rightly and fully adjuſted. So that I ſhall only proceed to make *Inferences* from them; which *Inferences*, in *this Part*, are all *affirmative.* Of theſe, the firſt is,

That theſe *Marine Bodyes* were born forth * of the *Sea* by the *Univerſal Deluge:* and that, upon the *Return* of the *Water* back again from off the *Earth*, they were left behind at *Land*. I.

* *Confer. Part* 3. *Sect.* 2. *Conſ.* 2, 3.

As this is a *Propoſition* of great *Weight* and *Conſequence*, I ſhall be very carefull and particular in the *Eſtabliſhment* of it; conferring every *Circumſtance*

Circumſtance of theſe *Marine Bodyes*, to ſee how they ſquare with it: and ſhall not diſmiſs it till I have evinced that *thoſe* which I preſt, in the *precedent Part*, as *Objections* againſt the *ſeveral Ways* there propounded, all fall in here, and are the cleareſt and moſt convincing *Arguments* of the *Truth* hereof: that *this*, and *this alone*, does naturaly and eaſily account for all thoſe *Circumſtances*: and fairly takes off all *Difficulties*.

Which being diſpatch'd, I return back to my *Obſervations*: and proceed upon *them* to repreſent the *Effects* that the *Deluge* had upon the *Earth*, and the *Alterations* that it wrought in the *Globe*; ſome whereof were indeed very *extraordinary*. Of which yet we have a plain and undeniable *Certainty*; the Evidences of them flowing directly and imdiately from the *Obſervations*: and being withall ſo full and clear that 'tis impoſſible theſe *Marine Bodyes* could have been any ways lodg'd in ſuch *Manner*, and to ſo great *Depths*, in the *Beds* of *Stone*, *Marble*, *Chalk*, and the reſt, had not

theſe

these Alterations all realy happen'd. Namely,

That during the *Time* of the *Deluge*, whilst the *Water* was *out* upon, and covered the *Terrestrial Globe*, all the *Stone* and *Marble* of the *Antediluvian Earth*: all the *Metalls* of it: all *Mineral Concretions*: and, in a Word, all *Fossils* whatever, that had before obtain'd any *Solidity*, were *totaly dissolved*, and their constituent *Corpuscles* all *disjoyned*, their *Cohæsion* perfectly ceasing. That the said *Corpuscles* of these *solid Fossils*, together with the *Corpuscles* of those which were not before *solid*, such as *Sand*, *Earth*, and the like: as also all *Animal Bodyes*, and *Parts* of *Animals*, *Bones*, *Teeth*, *Shells*: *Vegetables*, and *Parts* of *Vegetables*, *Trees*, *Shrubs*, *Herbs*: and, to be short, *all Bodyes* whatsoever that were either upon the *Earth*, or that *constituted* the *Mass* of it, if not quite down to the *Abyss* *, yet at least to the greatest *Depth* we ever dig; I say all these were *assumed up* promiscuously into the *Water*, and *sustained* in it, in such Manner that the *Water*, and *Bodyes* in it, together, 2.

* *Vid. Part* 3. *Sect.* I. *Consect.* I.

made up one common *confuſed Maſs.*

3. That at length all the *Maſs*, that was thus *born* up in the *Water*, was again *precipitated*, and *ſubſided* towards the Bottom. That this *Subſidence* happened generaly, and as near as poſſibly could be expected in ſo great a *Confuſion*, according to the *Laws* of *Gravity* *; *that* Matter, Body, or Bodyes, which had the *greateſt Quantity* or *Degree* of *Gravity*, ſubſiding *firſt* in Order, and falling *loweſt*: *that* which had the next, or a ſtill *leſſer Degree* of *Gravity*, ſubſiding *next* after, and *ſettling upon* the Precedent: and ſo on, in their *ſeveral Courſes*; *that* which had the *leaſt Gravity* ſinking not down till *laſt of all*, ſettling at the *Surface* of the *Sediment*, and *covering* all the *reſt*. That the *Matter*, ſubſiding thus, *formed* the *Strata* of *Stone*, of *Marble*, of *Cole*, of *Earth*, and the reſt; of which *Strata*, lying one upon another, the *Terreſtrial Globe*, or at leaſt as much of it as is ever diſplayed to view, doth mainly conſiſt. That the *Strata* being arranged in *this Order* meerly by

* Confer p. , &c.

by the *Disparity* of the *Matter*, of which each consisted, as to *Gravity*, that *Matter* which was *heaviest* descending *first*, and all that had the *same Degree* of *Gravity* subsiding at the *same Time* : and there being *Bodyes* of quite *different Kinds*, *Natures*, and *Constitutions*, that are nearly of the same *Specifick Gravity*, it thence happened that *Bodyes* of quite *different Kinds* subsided at the *same Instant*, fell together into, and composed the *same Stratum*. That for this Reason the *Shells* of those *Cockles*, *Escalops*, *Perewinkles*, and the rest, which have a *greater Degree* of *Gravity*, were enclos'd and lodg'd in the *Strata* of *Stone*, *Marble*, and the *heavier Kinds* of *Terrestrial Matter* ; the *lighter Shells* not not sinking down till *afterwards*, and so falling amongst the *lighter Matter*, such as *Chalk*, and the like, in all such *Parts* of the *Mass* where there happened to be any considerable Quantity of *Chalk*, or other *Matter lighter* than *Stone* ; but where there was *none*, the said *Shells* fell upon, or near unto, the *Surface*. And accordingly we *now* find the *lighter Kinds* of *Shells*, ſuch as thoſe of the *Echini*, and the like, very plentyfully in *Chalk*, but of the *heavier Kinds* ſcarcely one ever appears; *theſe* ſubſiding *ſooner*, and ſo ſettling *deeper*, and beneath the *Strata* of *Chalk*. That *Humane Bodyes*, the *Bodyes* of *Quadrupeds*, and other *Land-Animals*, of *Birds*, of *Fiſhes*, both of the *Cartilaginous*, *Squamoſe*, and *Cruſtaceous* Kinds: the *Bones*, *Teeth*, *Horns*, and other *Parts* of *Beaſts*, and of *Fiſhes*: the *Shells* of *Land-Snails*: and the *Shells* of thoſe *River* and *Sea Shell-fiſh* that were *lighter* than *Chalk*, *&c*: as alſo *Trees*, *Shrubs*, and all other *Vegetables*, and the *Seeds* of them: and that peculiar *Terreſtrial Matter* whereof *theſe* conſiſt, and out of which they are all *formed*; I ſay all *theſe* (except ſome *Mineral* or *Metallick Matter* happened to have been affixed to any of them *, whilſt they were *ſuſtained* together in the *Water*, ſo as to *augment* the *Weight* of them) being, Bulk for Bulk, *lighter* than *Sand*, *Marl*, *Chalk*, or the other ordinary *Matter* of the *Globe*, were not *precipitated* till the *laſt*, and ſo lay

* *As Pt.* 4. *Conſect.* 2.

lay *above* all the former, conſtituting the *ſupreme* or *outmoſt Stratum* of the *Globe.* That *theſe* being thus lodged *upon* the *reſt*, and conſequently more nearly *expoſed* to the *Air*, *Weather*, and other *Injuries*, the *Bodyes* of the *Animals* would ſuddenly *corrupt* and *rot*: the *Bones*, *Teeth*, and *Shells*, would likewiſe all *rot* in time, except thoſe which were ſecured by the extraordinary *Strength* and *Firmneſs* of their *Parts*, or which happened to be *lodged* in ſuch Places where there was great *Plenty* of *bituminous* or other like *Matter* to *preſerve*, and, as it were, embalm them: that the *Trees* would in Time alſo *decay* and *rot*, unleſs ſuch as chanced to be repoſed in, and ſecured by the ſame Kind of *Matter*: that the other more tender *Vegetables*, *Shrubs*, and *Herbs*, would *rot* likewiſe and *decay*. But the *Seeds* of all Kinds of *Vegetables*, being by this Means repos'd, and, as it were, *planted* near the *Surface* of the *Earth*, in a convenient and *natural Soil*, amongſt *Matter* proper for the *Formation* of *Vegetables*, would *germinate*, *grow* up, and repleniſh the

Face of the *Earth:* and that *vegetative Terrestrial Matter*, that fell, along with these, into this uppermost *Stratum*, and of which principaly it *consists*, hath been ever since, and will continue to be, the standing *Fund* and *Promptuary* out of which is derived the *Matter* of all *Animal* and *Vegetable Bodyes*, and whereinto, at the *Dissolution* of those *Bodyes*, that *Matter* is *restored* back again successively for the *Constitution* and *Formation* of others *.

* *Vid.* *Part* V. *Conf.* I.

4. That the *Strata* of *Marble*, of *Stone*, and of all other *solid Matter*, attained their *Solidity*, as soon as the *Sand*, or other Matter whereof they consist, was arriv'd at the *Bottom*, and well *settled* there. And that all those *Strata* which are *solid* at *this Day*, have been *so* ever *since* that *Time*.

5. That the said *Strata*, whether of Stone, of Chalk, of Cole, of Earth, or whatever other Matter they consisted of, lying thus each upon other, were all originaly *parallel:* that they were *plain*, *eaven*, and *regular:* and the *Surface* of the *Earth* likewise *eaven* and *spherical:* that they were *continuous*

continuous, and not *interrupted*, or *broken:* and the *whole Maſs* of the *Water* lay then *above* them all, and conſtituted a *fluid Sphere* environing the whole *Globe*.

That, after ſome time, the *Strata* were *broken*, on all Sides of the *Globe:* that they were *diſlocated*, and their Situation *varyed*, being *elevated* in ſome Places, and depreſs'd in others. 6.

That the *Agent*, or *Force*, which effected this *Diſruption*, and *Diſlocation* of the *Strata*, was ſeated *within* the *Earth*. 7.

That the *Irregularities* and *Inequalities* of the *Terreſtrial Globe* were *cauſed* by this *Means:* date their *Original* from *this Diſruption*, and are intirely owing unto it. That the natural *Grotto's* in Rocks, and thoſe *Intervalls* of the *Strata*, which, in my Obſervations, I call the *Perpendicular Fiſſures* *, are nothing but theſe *Interruptions* or *Breaches* of the *Strata*. That the more *eminent Parts* of the *Earth*, *Mountains* and *Rocks*, are only the *Elevations* of the *Strata*; theſe, wherever they were *ſolid*, rearing againſt and *ſupporting* each 8.

* *Confer Pag.* 11. *ſupra.*

 other

other in the *Posture* whereinto they were put by the *Bursting* or *Breaking up* of the *Sphere* of *Earth* *: and *not falling* down again, nor returning to their former and more level Site, as did the *Strata* of *Earth*, and other *Matter* that was *not solid*, and had no *Strata* of *Stone*, or other *consistent Matter*, interpos'd, amongst their *Strata* underneath, to uphold them in the *Posture* they were then raised into. For which Reason 'tis, that *Countryes* which abound with *Stone*, *Marble*, or other *solid Matter*, are *uneaven* and *mountainous*: and that those which afford none of these, but consist of *Clay*, *Gravel*, and the like, without any *Stone* interposed, are more *champaign*, *plain*, and *level*. That the *lower Parts* of the Earth, *Vallyes*, the *Chanel of the Sea*, and the rest, are nothing but *Depressions* of the *Strata*. That *Islands* were form'd and distinguish'd by the *Depression* or *sinking down* of the *Strata* lying betwixt each of them, and betwixt them and the *Continent*. In one Word, that the *whole Terraqueous Globe* was, by *this Means*, at the *Time* of the *Deluge*, put

* Confer *Consect.* 6. supra.

put into the *Condition* that we *now* behold.

Here was, we ſee, a mighty *Revolution:* and *that* too attended with *Accidents* very ſtrange and amazing: the moſt horrible and portentous *Cataſtrophe* that *Nature* ever yet ſaw: an elegant, orderly, and habitable *Earth* quite *unhinged*, ſhatter'd all to *Pieces*, and turned into an Heap of *Ruins:* Convulſions ſo exorbitant and unruly: a *Change* ſo exceeding great and violent, that the very *Repreſentation* alone is enough to ſtartle and ſhock a Man. In Truth the Thing, at firſt, appear'd ſo *wonderful* and *ſurprizing* to me, that I muſt confeſs I was for ſome Time at a Stand. Nor could I bring over my *Reaſon* to aſſent, untill, by a deliberate and careful *Examination* of all *Circumſtances* of theſe *Marine Bodyes*, I was abundantly convinced that they could not have come into thoſe *Circumſtances* by any other *Means* than ſuch a *Diſſolution* of the *Earth*, and *Confuſion* of *Things*. And were it not that the *Obſervations*, made in ſo *many*, and thoſe ſo *diſtant Places*, and repeated ſo often

ten with the moſt ſcrupulous and diffident *Circumſpection*, did ſo eſtabliſh and aſcertain the Thing, as not to leave any Room for *Conteſt* or *Doubt*, I could ſcarcely ever have credited it.

And though the whole *Series* of this extraordinary *Turn* may ſeem at firſt View to exhibit nothing but *Tumult* and *Diſorder*: nothing but Hurry, Jarring, and Diſtraction of Things: though it may carry along with it ſome ſlight Shew that 'twas managed *blindly* and at *random*; yet if we draw ſomewhat nearer, and take a cloſer Proſpect of it: if we look into its *retired Movements*, and more ſecret and latent *Springs*, we may there trace out a *ſteady Hand*, producing *Good* out of *Evil*: the moſt conſummate and abſolute *Order* and *Beauty*, out of the higheſt *Confuſion* and *Deformity*: acting with the moſt exquiſite *Contrivance* and *Wiſdom*: attending vigilantly throughout the whole *Courſe* of this grand *Affair*, and directing all the ſeveral *Steps* and *Periods* of it to *an End*, and that a moſt *noble* and *excellent* one; no leſs than the *Happineſs* of the whole

whole Race of *Mankind*: the *Benefit*, and universal *Good*, of all the many *Generations* of *Men* which were to come after: which were to inhabit *this Earth*, thus *moduled anew*, thus suited to their *present Condition* and *Necessities*.

But the *Presidence* of that *mighty Power* in this *Revolution*: its particular *Agency* and *Concern* therein: and its *Purpose* and *Design* in the several *Accidents* of it, will more evidently appear, when I shall have proved,

That, altho' one *Intention* of the *Deluge* was to inflict a deserved *Punishment* upon *that Race* of *Men*, yet it was not *soley-levell'd* against *Mankind*, but *principaly* against the *Earth* that then was; with Design to *destroy* and *alter* that *Constitution* of it, which was apparently *calculated* and *contrived* for a *State* of *Innocence*: to fashion it afresh, and give it a *Constitution* more nearly accommodated to the *present Frailties* of its *Inhabitants*.

That the said *Earth*, though not indifferently and alike *fertil* in all Parts of it, was yet generaly much

more

more fertil than *ours* is. That the *exteriour Stratum* or *Surface** of it, consisted entirely of a kind of *terrestrial Matter* proper for the *Nourishment* and *Formation* of *Plants*, and this in great *Plenty* and *Purity*; being little, or not at all, entangled with an *Intermixture* of meer *Mineral Matter* that was unfit for *Vegetation*. That its *Soil* was more *luxuriant*: and teemed forth its *Productions* in far greater *Plenty* and *Abundance* than the *present Earth* does.* That the *Plough* was *then* of no Use, and not invented 'till *after* the *Deluge*; *that Earth* requiring little or *no Care* or *Culture*, but yielding its Increase *freely*, and without any considerable *Labour* and *Toil*, or Assistance of *Humane Industry*; by this Means allowing *Mankind* that *Time*, which must otherwise have been spent in *Agriculture*, Plowing, Sowing, and the like, to far more divine and noble *Uses*: to *Purposes* more agreeable to the *Design* of their *Creation*; there being no Hazard, whilst they continu'd in that *State* of *Perfection*, of their Abusing this *Plenty*, or Perverting it to any other *End*

* *Confer pag.* 13. *supra*: and *Part* V. *Confect.* 1.

* *Confer Part* 6. *Confect.* 5.

End than the *Suſtenance* of *Nature*, and the neceſſary *Support* of *Life*.

That when *Man* was *fallen*, and had abandon'd his *primitive Innocence*, the *Caſe* was much *alter'd*: and a far *different Scene* of Things preſented. That generous *Vertue*, maſculine *Bravery*, and prudent *Circumſpection* which he was before Maſter of, now deſerted him, together with that *Innocence* which was the *Baſis* and *Support* of all: and a ſtrange *Imbecility* immediately ſeiz'd and laid hold of him. He became *puſillanimous*, and was eaſily ruffled with every little *Paſſion* within: ſupine, and as openly expoſed to any *Temptation* or *Aſſault* from without. And *now* thoſe *exuberant Productions* of the *Earth* became a continual *Decoy* and *Snare* unto him. They only excited and fomented his *Luſts*: and miniſtred plentyfull *Fewel* to his *Vices* and *Luxury*. And the *Earth* requiring little or no *Tillage*, there was little Occaſion for *Labour*; ſo that almoſt his *whole Time* lay upon his Hands, and gave him *Lieſure* to contrive, and full ſwing to purſue his *Follies*; by which

which *Means* he was laid open to
all manner of *Pravity*, *Corruption*,
and *Enormity*. So that we need not
be much ſurprized to hear *That the*
Wickedneſs of Man was great in the
Earth, *and that every Imagination of*
the Thoughts of his Heart was only
* *Gen*.vi.5. *evil continualy**: nor more, that
that Generation of *Men* was more
particularly addicted to *Intemperance*,
Senſuality, and *Unchaſtity*: that they
ſpent their Time in Gluttony, in
Eating and Drinking, in Luſt and
Wantonneſs, or, as the ſacred Writer cleanly and modeſtly expreſſes
it, in *Marrying*, *and giveing in Mar-*
* *Matth.* *riage**, and this without Diſcretion
xxiv, 38. or Decency, without Regard to Age
or Affinity, but promiſcuouſly, and
with no better a Guide than the
Impulſes of a brutal Appetite, *They*
took them Wives of all which they
* *Gen*.vi.2. *choſe**; Plenty and Abundance,
Idleneſs and Eaſe, ſo naturaly *che-*
riſhing and *promoting* thoſe *particular*
Vices. Nor laſtly, is it ſtrange that
the *Apoſtacy* was ſo *great*, the *In-*
fection ſo *univerſal*: that *the Earth*
was filled with Violence, and that *all*
* *Gen*. vi. *Fleſh had corrupted his Way**; the
11, 12.

Cauſe

Cauſe of this *Corruption*, the *Fertility* of *that Earth*, being ſo *univerſal*, ſo *diffuſive* and *epidemical*. And indeed, 'twould be very hard to aſſign any *other* ſingle *Cauſe*, beſides *this*, that could ever poſſibly have had ſo *ſpreading* and *general* an *Effect* as *this* had. The *Pravity* of *humane Nature* is not, I fear, *leſs* than it was *then*: the *Paſſions* of *Men* are yet as *exorbitant*, and their *Inclinations* as *vicious*: Men have been *wicked* ſince the *Deluge*: they are ſo *ſtill*: and *will* be ſo, but not *univerſaly*. There are now *Bounds* ſet to the *Contagion*: and 'tis reſtrained by removing the *main Cauſe* of it. But *there*, the *Venom* manifeſted it ſelf on all Hands: *ſpread* far and near, ſcarcely ſtopping 'till 'twas inſinuated into the whole *Maſs* of *Mankind*: and the *World* was little better that a common Fold of *Phreneticks* and *Bedlams*.

That to *reclaim* and *retrieve* the *World* out of this *wretched* and *forlorn State*, the common *Father* and *Benefactor* of *Mankind* ſeaſonably interpoſed his *Hand*: and reſcued miſerable *Man* out of the groſs *Stupidity*

pidity and *Senſuality* whereinto he was thus unfortunately plunged. And this was effected partly by tying up his Hands, and *ſhortning* the *Power* of *Sinning:* checking him, in the Career of his *Follies*, by *Diſeaſes* and *Pains:* and ſetting *Death*, the *King* of *Terrors*, which before ſtood *aloof off*, and at the *long Diſtance* of eight or nine hundred Years, now much *nearer* to his View, ordaining that *his Days ſhall be* but
† Gen. vi. 3. *an hundred and twenty Years* †: and partly by *Removing* the *Temptation*, and *Cauſe* of the *Sin:* by *Deſtroying** that *Earth* which had furniſh'd forth *Maintenance* in ſuch *Store* unto it: by *changing* that *Conſtitution* of it, and

* Gen. vi. 13. *And behold I will DESTROY them with THE EARTH.* And again, at the Covenant made with *Noah*, after the Deluge, more diſtinctly, *Gen.* ix. 11. *Neither ſhall all Fleſh be cut off any more by the Waters of a Flood: neither ſhall there any more be A FLOOD TO DESTROY THE EARTH*; the latter Part whereof is render'd ſomewhat more expreſly by the Septuagint ϗ οὐκ ἔτι ἔςαι κατακλυσμὸς ὕδατος καταφθεῖραι ΠΑ͂ΣΑΝ τὴν γῆν. i. e. *And there ſhall not be any more a Deluge of Water to deſtroy the WHOLE EARTH.* The vulg. Lat. hath it, *Neque erit deinceps Diluvium* diſſipans *terram*, i. e. *Neither ſhall there be hereafter a Deluge to* diſſipate [or diſſolve] *the Earth.* And of this *Diſſolution* of the Earth there was a *Tradition* amongſt the *Antients*, both *Jews* and *Gentiles*. *Vid. Part.* 3. *infra*, near the End.

and rendring it more *agreeable* to the *laps'd* and *frail State* of *Mankind.* That *this Change* was not wrought by altering either the *Form* of the *Earth*, or its *Position* in respect of the *Sun*, as was not long ago surmised by a Learned Man *, but by *Dissolving* † it: by *Reducing* all the *Matter* of it to its *first constituent Principles*: by *Mingling*, and *Confounding* them, the *Vegetative* with *mineral Matter*, and the *different Kinds* of *mineral Matter* with *each other* *: and by *Retrenching* a considerable Quantity of the *vegetable Matter*, (which lay in such *Plenty* and *Purity* at the *Surface* of the *Antediluvian Earth*, and rendred it so exuberantly *fruitful*) and *Precipitating* it, (at the Time of the *Subsidence* * of the general Mass of *Earth* and other *Bodyes*, which were before *raised* up into the *Water*) to such a *Depth* as to *bury* it, leaving only so much of it near the *Surface* as might just sufficiently *satisfy* the *Wants* of *humane Nature*, but little or no more; and even *that* not *pure*, not free from the Intermixture of meer *steril* mineral *Matter*, and such as is

* Dr. Burnet *Theory of the Earth.*

† *Vid. Consect.* 2. *supra.*

* *Part* 4. *Consect.* 3.

* *Vid. Consect.* 3. *supra.*

H in

in no wiſe fit for the *Nutrition* of *Vegetables*; but ſo that it ſhould require *Induſtry* and *Labour* to excite it, and not yield a competent *Crop* without *Tillage* and *Manure*. That, by this *Means*, a great Part of that *Time*, which the *Inhabitants* of the *former Earth* had to *ſpare*, and whereof they made ſo *ill Uſe*, was *employ'd*, and *taken up* in *Digging* and *Plowing*, in making Proviſion for *Bread*, and for the *Neceſſities* of *Life*: and that *Exceſs* of *Fertility*, which contributed ſo much to *their Miſcarriages*, was *retracted* and *cut off*.

That had the *Deluge* been aimed *only at Mankind*, and its utmoſt *Deſign* meerly to *puniſh* that *Generation*, and thereby to deterr *Poſterity* from the like *Offenſes*, *this* might have been brought about by *Means* much more *compendious*, and *obvious* too, and yet equaly *terrifying* and *exemplary*. *Mankind*, I ſay, might have been *taken off* at a far *cheaper Rate*; without this *Ranſacking* of *Nature*, and Turning all Things topſie-turvy: without this *Battering* of the *Earth*, and *Unhinging* the whole *Frame* of the *Globe*. The Buſineſs might

might have been done as effectualy by *Wars*; the *Heart* of every *Man* of them was in the *Hand* of *God*, and he could easily have made them *Executioners* of his *Wrath* upon *one another*. He had the Command of *Famine*, of *Pestilence*, and a thousand other *Disasters*, whereby he could have carry'd them off by *Sholes*, yea swept them all clear away. Besides, he had the whole *Artillery* of the *Sky* in his *Power*, and might presently have *Thunder-struck* them all, or *destroyed* them by *Fire* from *Heaven*. But none of all *these* were used; though 'tis most apparent that any of them would have been as *fatal* and *pernicious* to *Man* as the *Deluge* was. For the *Design* lay a great deal *deeper*: and *these* would have fallen short of it. *These* would never have reach'd the *Earth*: nor affected *that* in the least. They could never have touch'd the *Head*: or stopped the *Source* of these unhappy *Misdemeanours*, for which the *Punishment* was sent. *That* was what nothing but a *Deluge* could reach; and as long as the *Cause* remained: as long as the *old Tempta-*

tion was ſtill *behind*, every *Age* would have lain under *freſh Inducements* to the *ſame Crimes*: and there would have been a *new Neceſſity* to *Puniſh* and *Reclaim* the *World*: to *Depopulate* the *Earth*, and *Reduce* it again to a vaſt *Solitude*, as conſtantly as there ſucceeded a *new Age* and *Race* of *Men*. For the *Terror* of the *Calamity* would not have extended it ſelf much *farther* than the *Men* which *ſuffer'd* under the Weight of it: and a *few Years* would have worn out, in great meaſure, the *Impreſſions* it made. This we ſee even from the *Example* of the *Deluge* it ſelf. As formidable as *that* was to thoſe who lived at, or near the *Time* of it, who ſaw the prodigious *Devaſtations* it had made, the *horrible Methods* by which 'twas brought about, and the *Reaſon* why 'twas inflicted: and to *their Poſterity*, for a few *Generations*; the *Fright* was not *laſting*: 'twas not *long* e'er the *Sting* of it was worn out. And though the *Elder Ages* knew full well that there had been ſuch a *Deluge*: and had ſome *Tradition* of the cruel *Deſolation* it made; yet by

Degrees

Degrees the *Particulars* of it were *drop'd*, and the moſt *frightful Paſſages* bore the *leaſt Share* in the *Relation*; being probably ſo *ſtrange* as to be hardly *credible*: and carrying rather an Appearance of *Figment* and *Invention*, in *thoſe* that handed down the *Memory* of it, than of *Truth* and *Reality*. So that upon the whole tis very plain that the *Deluge* was not ſent only as an *Executioner* to *Mankind*: but that its *prime Errand* was to *Re-form* and *New mold* the *Earth*.

That therefore, as much *Harſhneſs* and *Cruelty* as this *great Deſtruction* of *Mankind* ſeemingly carryes along with it: as *wild* and *extravagant* a Thing as that *Diſſolution* of the *primitive Earth* appear'd at firſt Sight; all the *Severity* lay in the *Puniſhment* of *that Generation*, (which yet was no more than what was *highly juſt*, yea and *neceſſary* too:) and the *whole* of the *Tragedy* terminated there. For the *Deſtruction* of the *Earth* was not only an Act of the profoundeſt *Wiſdom* and *Forecaſt*, but the moſt monumental *Proof*, that could ever poſſibly have been given, of *Goodneſs*, *Compaſſion*, and *Tenderneſs*, in the *Author* of our *Being*; and *this* ſo *liberal* too and *extenſive*, as to reach all the *ſucceeding Ages* of *Mankind:* all the *Poſterity* of *Noah:* all that ſhould dwell upon the thus *renewed Earth* to the *End* of the *World*; by this *Means* removing the old *Charm:* the *Bait* that had ſo long *bewildred* and *deluded* unhappy *Man:* ſetting him once more upon his *Legs:* reducing him from the moſt abject and ſtupid *Ferity*, to his *Senſes*, and to *ſober Reaſon:* from the moſt deplorable *Miſery* and *Slavery*, to a *Capacity* of being *happy*.

That *this Remedy* was made uſe of notwithſtanding that it ſtruck as deep at the *Intellectuals*, as at the *Morals* of *Mankind:* that *Ignorance* and *Rudeneſs* would be as neceſſary a *Conſequence* of it, as *Reformation* of *Life:* and that this ſo general *Employ* and *Expenſe* of their *Time* would as aſſuredly *curtail* and *retrench* the ordinary Means of *Knowledge* and *Erudition**, as 'twould *ſhorten* the *Opportunities* of *Vice*. And ſo accordingly it fell out; an univerſal

* *Confer p.* 59. & *ſeqq.*

universal *Rusticity* presently took place: spread on all Hands, and stop'd not till it had over-run the *whole Stock* of *Mankind.* Those *first Ages* of the *new World* were *simple* and *illiterate* to Admiration, and 'twas a *long Time* e'er the *Cloud* was withdrawn: e'er the least Spark of *Learning* (I had almost said of *Humanity*) broke forth, or any Man betook himself to the *Promoteing* of *Science.* Nay the *Effects* of it are visible to this Hour: a general *Darkness* yet prevails, and hangs over *whole Nations:* yea the far greater Part of the *World* is *still barbarous* and *savage.* I say, tho' 'twas most evident that *this Remedy* must needs have *this Consequence* also as well as the *other*, yet it was not *suspended* or chang'd upon that Account; an egregious and pregnant *Instance* how far *Vertue* surpasses *Ingenuity:* how much an honest *Simplicity*, *Probity* of *Mind*, *Integrity* and *Incorruptness* of *Manners*, is preferable to *fine Parts*, profound *Knowledge*, and subtile *Speculations.* I would not have *this* interpreted an *Invective* against *Humane Learning*, or a decrying any commendable *Accompliſhments* either of *Body* or *Mind*, (that is what no Man will, I hope, ſuſpect me of) but only an Intimation that *theſe* are not of any ſolid *Uſe*, or real *Advantage*, unleſs when aiding and ſerviceable to the *other*.

Nor does this *grand Cataſtrophe* only preſent us with Demonſtrations of the *Goodneſs*, but alſo of the *Wiſdom* and *Contrivance* of its *Author*. There runs a *long Train* of *Providence* thro' the *whole* : and ſhines brightly forth of all the various *Accidents* of it. The *Conſolidation* of the *Marble*, and of the *Stone*, immediately after their *Settlement* to the Bottom : the *Diſruption* of the *Strata* afterwards : their *Diſlocation*, the *Elevation* of ſome, and *Depreſſion* of others of them, did not fall out at *Random*, or by *Chance* : but were managed and directed by a more ſteady and diſcerning *Principle*. For Proof whereof, this is indeed the proper *Place* ; but, in regard that there are ſome Things advanced in the ſucceeding, or *third Part* of this *Diſcourſe*, which give ſome *farther Light* to this Matter,

ter, I ſhall beg leave to break off here, and to deferr it a while, untill I have firſt propoſed them.

Thus have I drawn up a *brief Scheme* of what befell the *Earth* at the *Deluge:* and of the *Change* that it then underwent. I have, by comparing its *Antediluvian* * with its *preſent State*, found where chiefly the *Difference* lay; *viz.* in *Degree* of *Fertility.* I have endeavour'd alſo to diſcover the *Reaſon* why this *Change* was made in it. For, ſince that the *Proceſs* of it was ſo *ſolemn* and *extraordinary:* that there were ſo *many*, and thoſe ſo *ſtrange Things* done: that the *firſt Earth* was perfectly *unmade* again, taken all to Pieces, and *framed a-new*; and, indeed, the very *ſame Method* that was uſed in the *original Formation* of it, uſed likewiſe in this *Renovation*; *our Earth* ſtanding the *firſt Step* after its *Diſſolution*, in the *ſame Poſture* that the *Primitive Earth* did the firſt Step after its *Riſe* out of Nothing; which the *Reader* will eaſily find by conferring the *fifth Propoſition* of *this Part* with *Gen.* i. *ver.* 2. and 9: ſince likewiſe there

* *Conf. Part* 6.

was

* *Confer Part* 3. *Sect.* 2. *Conſ.* 7.

was ſo *mighty* an *Hand** concern'd, and which does not act without *great* and *weighty Reaſons*, there could be no *Doubt* but that there was ſome real and very neceſſary *Cauſe* for the making that *Alteration*. Nor was ſuch a *Cauſe* very hard to be found out. The *firſt Earth* was ſuited to the *firſt State* of *Mankind*, who were the *Inhabitants* of it, and for whoſe *Uſe* 'twas made. But when *Humane Nature* had, by *the Fall*, ſuffer'd ſo great a *Change*, 'twas but neceſſary that the *Earth* ſhould undergo a *Change* too, the better to *accommodate* it to the *Condition* that *Mankind* was *then* in: and *ſuch* a *Change* the *Deluge* brought to paſs.

But leaſt the *Brevity* which I have above uſed, and which indeed I am ty'd up to, in my Repreſentation of this Matter, ſhould render it liable to *Miſconſtruction*: or that any one ſhould ſuſpect, that what I have deliver'd concerning the *Fertility* of *that Earth*, does not well ſquare with the *Moſaick Deſcription* of it, I muſt beg leave to make a *Digreſſion* here, that I may explain

my

my ſelf a little more upon that Head. And that the *Reader* may himſelf be *Judge* in the Caſe, I ſhall fairly lay down *Moſes*'s Senſe of it in his own Words.

Verſ. 17. *And unto Adam he ſaid, be-* Gen. iii.
cauſe thou haſt hearkned unto the Voice of thy Wife, and haſt eaten of the Tree of which I commanded thee, ſaying, thou ſhalt not eat of it, Curſed is the Ground for thy ſake; in Sorrow ſhalt thou eat of it all the Days of thy Life.

Verſ. 18. *Thorns alſo and Thiſtles ſhall it bring forth to thee: and thou ſhalt eat the Herb of the Field.*

Verſ. 19. *In the Sweat of thy Face ſhalt thou eat Bread, till thou return unto the Ground; for out of it waſt thou taken: for Duſt thou art, and unto Duſt ſhalt thou return.*

Verſ. 23. *Therefore the Lord God ſent him forth from the Garden of Eden, to till the Ground from whence he was taken.*

Verſ. 2. *Cain was a Tiller of the* Gen. iv.
Ground.

All which may be well reduced to two plain and ſhort *Propoſitions.*

1. That

1. That *Adam*'s Revolt drew down a *Curſe* upon the *Earth*.

2. That there was ſome ſort of *Tillage*, or *Agriculture* uſed before the Deluge.

As to the former, the *Curſe* upon the *Earth*, I ſhall not in the leaſt go about to extenuate the Latitude of it: or to ſtint it only to the Production of Weeds, of *Thorns*, *Thiſtles*, and other the leſs uſeful Kinds of *Plants*: but ſhall give it its full Scope, and grant that no leſs than an univerſal *Reſtraint* and *Diminution* of the *primitive Fruitfulneſs* of the *Earth* was intended by it; this indeed ſeeming to be the plain and genuine *Meaning* of the *Words*. But the Queſtion is, whether this *Curſe* was *preſently* inflicted or not: whether it was ſucceeded with an *univerſal Sterility*, and the *Earth*'s native and original *Exuberance* all ſtraitways check'd and turned to as general a *Deſolation* and *Barrenneſs*. And here I entreat it may be taken Notice, that *this* was but *one*, and that much the *leſſer* Part of the *Sentence* paſs'd upon *Adam*. The *other* was *Death* *, which, 'tis moſt certain,

* *Confer* Rom. v. 12. *and* 1 Cor. xv. 21, 22.

certain, was not *immediately* inflicted. And yet *this* was pronounc'd at the ſame Time, and with the ſame Breath, that the other was; *Unto Duſt ſhalt thou return.* Nay and much more emphaticaly a little before*, *In the Day that thou eateſt thereof thou ſhalt* ſurely *dye.* This was exceedingly peremptory: and the *very Day* fix'd likewiſe. Notwithſtanding, through the Clemency and Goodneſs of God, *Execution* was delay'd for a long time; *Adam* being *repriev'd* for eight or nine hundred Years*. The Dominion of *Death* over him commenc'd indeed not only the ſame Day that *Sentence* was paſt, but the very Minute that he taſted the forbidden *Fruit*: and *Mortality* went hand in hand with the *Tranſgreſſion.* But 'twas a *long Time* before it had rais'd any Trophies: or made a final and abſolute Conqueſt.

* Gen. ii. 17.

* Gen. v. 5.

Why therefore may we not as well ſuppoſe the *other Part* of the *Sentence*, the *Sterilizing* the *Earth*, was alſo *ſuſpended* for ſome Time, and *deferr'd* 'till the *Deluge* happen'd, and became the Executioner of it? 'Tis certainly very hard to imagine that

that God ſhould *deſtroy* the *Work* of his Hands almoſt as ſoon as he had *finiſh'd* it: that all Things ſhould be *unhinged* again by ſuch time as they were well *ranged* and put in *Order*: and that the *Fragrancy* and lovely *Verdure* which then appear'd every where, and which had but juſt ſhew'd it ſelf, ſhould be nipp'd in the Bud, and *blaſted* all of a ſudden. To be ſhort, 'tis, I think, moſt apparent, that as on the other Part *Mortality* did preſently *enter* and *take place*, but got not *full Poſſeſſion* of *many Ages* after: ſo here, *Thorns*, *Thiſtles*, and other the like *Conſequences* of this *Curſe*, immediately ſprung out of the *Ground*, and manifeſted themſelves on every Side; but it had not its full *Effect*, nor was the Earth *impoveriſh'd*, or its *Fertility* ſenſibly *curb'd*, 'till the Deluge. And, for Proof of *this*, I appeal to the *Remains* of *that Earth*; the *Animal* and *Vegetable Productions* of it ſtill preſerv'd; the vaſt and incredible *Numbers* whereof notoriouſly teſtify the extreme *Luxuriance* and *Fœcundity* of it *. And I need but produce *theſe* as Evidences

* *Confer* Part VI. *Conſ.* 4, 5.

Evidences that at the Time that the *Deluge* came, the *Earth* was so *loaded* with *Herbage*, and *throng'd* with *Animals*, that such an *Expedient* was even wanting to ease it of the *Burden*, and to make Room for a *Succession* of its *Productions.* For *this* also I appeal to *Moses* himself, who openly acknowledgeth that this *Curse* did not take Place effectualy 'till the *Deluge.* For he tells us, that, after the *Deluge* was over, and *Noah* and his Family came forth of the Ark, *He builded an Altar unto the Lord, and offer'd Burnt-Offerings on the Altar: and the Lord smelled a sweet Savour, and the Lord said in his heart, I will not again* CURSE THE GROUND *any more, neither will I again smite any more every thing living as I have done**. Wherein he plainly refers to the *Curse* denounc'd above, at the Apostacy of *Adam*; implying that it was not *fulfilled* 'till the *Deluge.* And, a little after, he as plainly intimates, that the *Fulfilling* of it lay in the *Destruction* of the *Earth* then wrought. For, speaking again of the same Thing, instead of the Expression

* *Gen.* viii. 20, 21.

preſſion [*Curſe the Ground*] here uſed, he makes uſe of [*Deſtroy the Earth.*] The whole Paſſage runs thus; *And I will eſtabliſh my Covenant with you, neither ſhall all Fleſh be cut off any more by the Waters of a Flood: neither ſhall there any more be a* FLOOD TO DESTROY THE EARTH *.

* Gen. ix. 11.

Nor is it indeed in any wiſe ſtrange that this *Curſe* had not its *Effect* ſooner; eſpecialy ſince 'twas not limited to any Time. There are ſo many *Precedents* on Record in *Holy Writ* of this Way of proceeding, that no one can be well ignorant of them; ſo that I ſhall not need to charge this Place with more than one, and that ſhall be the Caſe of *Ham*; for which we are likewiſe beholden to the ſame Author, *Moſes*. This Perſon, by his indiſcreet and unnatural *Derideing* and *Expoſing* of his *Father*, incurrs his *Indignation*, and *Curſe*. But, which is very remarkable, *Noah* does not lay the *Curſe* upon *Ham*, who was actualy guilty of the Crime; whether out of greater *Tenderneſs*, he being of the two nearer related unto him, or for

for what other Reaſon I ſhall not here enquire, but transfers it to *Canaan. Curſed be Canaan, a Servant of Servants ſhall he be to his Brethren: to Shem* and to *Japhet.* Gen. ix. 25, 26, 27. Nay, which is ſtill more, this was never inflicted upon *Canaan* in *Perſon,* but upon his *Poſterity*; and that not 'till many Generations afterwards, at ſuch Time as the *Iſraelites,* returning out of *Egypt,* poſſeſs'd themſelves of the Country of the *Canaanites,* and made *them* their *Servants.* The Story is ſo well known, that I ſhall not need to point it out to the Reader, who may peruſe it at his Leiſure. 'Twas well onwards of a thouſand Years before ever this *Curſe* began to take effect: before the *Canaanites* were brought under *Servitude* by the *Iſraelites,* who were deſcended from *Shem:* and a great many more before 'twas finaly *accompliſh'd,* and *they* ſubjected unto the Poſterity of *Japhet.* To conclude, 'twas realy a *longer time* before *this,* than it was before *the other,* the *Curſe* upon the *Earth,* was fully brought about.

To proceed therefore to the *other Point*, the *Tillage* of the *Earth* before the *Deluge*. That there was *Tillage* beſtow'd upon it *Moſes* does indeed intimate in general and at large; but whether it was beſtow'd on *all*, or only upon *ſome Parts* of *that Earth*: as alſo *what Sort* of *Tillage* that was, and what *Labour* it coſt, is not expreſs'd; ſo that for all this we are at *Liberty*, and may uſe our *Diſcretion*. For the preſent I muſt paſs by the *Enquiry*: but in due Place I hope to give ſome Satisfaction in it, and to ſhew that *their Agriculture* was nothing near ſo *laborious*, and *troubleſome*, nor did it take up ſo much *Time* as *ours* doth. That's a *Conſequence* of the *Proof* of the greater *Fertility* of that *Earth*; it being plain that the more it exerted that *Fertility*, the leſs Need there was of *Manure*, of *Culture*, or *Humane Induſtry* to excite and promote it. Nor can any Man reaſonably ſuſpect, becauſe of this Mention of *Tillage*, that the *Curſe* upon the *Ground* was come on, or that the *primitive Exuberance* of the *Earth* was *leſſen'd* and *abridg'd*, before

fore the *Deluge*; for *Moses* makes mention of *Tillage* before ever *Adam* was created; *There was not*, says he *, *a man to Till the Ground*: and consequently, there would have been requisite such a *Tillage*, as *this* which he speaks of in these three Chapters, though the *Curse* had never been *denounc'd*, or *Man* had not *fallen*. But 'tis highly probable that upon *Adam*'s Disobedience, Almighty God chased him out of *Paradise*, the *fairest* and most *delicious* Part of *that Earth*, into some other the most *barren* and *unpleasant* of all the whole *Globe*; the more effectualy to signify his *Displeasure*, and to convince that unhappy Man how great a *Misfortune* and *Forfeiture* he had incurred by his late *Offense*. And here, above all other Parts of the *Earth*, there would be *Work* and *Employ* for him, and for his Son *Cain*.

* Gen. ii. 5.

And thus much may serve, for the present, to shew that my *Account* of the *Antediluvian Earth* is so far from *Interfereing* with that which *Moses* hath given us, that it holds forth a natural and unforc'd *Inter-*

pretation of his *Senſe* on this *Subject*. There are a few other *Paſſages* in the ſame *Author* which may require ſome *Explication*; but they are none of them *ſuch* that a Reader of moderate Underſtanding may not eaſyly clear, without my Aſſiſtance, ſo that I ſhall not crowd *this Piece* with them; for I fear 'twill be thought that I have already taken too great a *Liberty*.

The *Compaſs* that I am confin'd unto, by *the Rules* of this kind of *Writing*, is ſo *narrow*, that I am forced to paſs over *many Things* in Silence, and can but juſt touch upon *others*. To lay down every thing at *Length*, and in its *full Light*, ſo as to obviate all *Exceptions*, and remove every *Difficulty*, would carry me out too far beyond the *Meaſures* allowed to a *Tract* of *this Nature*. That's the *Buſineſs* of the *Larger Work*, of which *this* is only the *Module* or *Platform*. In *that Work* I hope to make amends for theſe *Omiſſions*, and particularly ſhall conſider

What was the immediate *Inſtrument* or *Means* whereby the *Stone*, and

and other *ſolid Matter* of the *Antediluvian Earth* was *diſſolved*, and reduced to the *Condition* mentioned *Conſect.* 2. of this Part.

Why the *Shells*, *Teeth*, *Bones*, and other Parts of *Animal Bodyes*: as alſo the Trunks, Roots, and other Parts of *Vegetables*, were not *diſſolved*, as well as the *Stone*, and other *Mineral Solids* of *that Earth*. Of this I ſhall aſſign a plain and *Phyſical Reaſon*, taken meerly from the *Cauſe* of the *Solidity* of theſe *Mineral Bodyes*; which I ſhew to be quite *different* from *that* whereunto *Vegetables* and *Animals* owe the *Cohæſion* of their *Parts*: and that *this* was *ſuſpended* at the Time that the *Water* of the *Deluge* came forth; which the *other* (I mean the *Cauſe* of the *Cohæſion* of the Parts of *Animals* and *Vegetables*) was not.

What was the Reaſon that (in caſe the *Terreſtrial Globe* was entirely *diſſolved*, and there be now, and was then, a *Space* or *Cavity*, in the *Central* Parts of it, ſo *large* as to give Reception to that mighty *Maſs* of *Water* which covered the *Earth* at the *Deluge**) the *Terreſtrial Matter*

* *Vid. Part* 3. *Sect.* 1. *Conſect.* 1. *and Sect.* 2. *Conſ.* 2. 3.

ter which *first subsided*, (as in *Consect.* 3. *supra*) did not *fill* the said *Cavity*, and *descend* quite down to the *Center*, but *stop'd* at that *Distance* from it, forming an arched *Expansum*, or rather a *Sphere* around it; which is now the lowest *Stratum*, and *Boundary* of that vast *Receptacle* of *Water*. As also how *this Water* was *raised* at the *Deluge*: by what *Issues* or *Outlets* it came forth: what *succeeded* into the *Room* of it whilst absent: and *which Way* it *return'd* back again.

By what Means the *Strata* of *Stone*, and *Marble*, acquir'd such a *Solidity*, as soon as the *Matter*, whereof they consist, had *subsided*, and was well *settled* to the Bottom, as in *Consect.* 4. of this Part.

What was the immediate *Agent* which effected that *Disruption* of the *Strata*, and their *Dislocation* afterwards; whereof in *Consect.* 6. of this Part.

And because there have been some *Conjectures* formerly started by *Learned Men* about the *Formation* of *Sand-Stone*, the *Origin* of *Mountains*, and of *Islands*, that are *repugnant*

to

to what I have *here* advanc'd upon thoſe Subjects, I am obliged particularly to conſider them. That therefore they may not remain as *Obſtacles* to thoſe who are leſs ſkilfull in theſe Things, I ſhall weigh their *Arguments*, detect the *Invalidity* of them, and prove, againſt them,

That the *Sand-Stone*, now in *Being*, is not as old as the *Earth* it ſelf: nor hath it been *conſolidated* ever ſince the *Creation* of the *World*, as ſome Authors have believed.

That *Sand-Stone* does not now *grow* by *Juxtapoſition*, as they ſpeak; that is by continual *Addition* of *new Matter*; in like Manner as the Bodyes of *Animals* and of *Vegetables* grow, and are augmented; as others were of Opinion.

That *Sand-Stone* does not *ſtill conſolidate*: *i. e.* that *Matter* which was, a few Years ago, *lax*, *incoherent*, and in *Form* of *Earth*, or of *Sand*, does not become daily more *hard* and *conſiſtent*, and by little and little acquire a perfect *Solidity*, ſo as to turn to *Stone*; as others have aſſerted.

That the *Mountains* of *our Earth* have not had *Being* ever ſince the *Creation:* and ſtood as long as the *Earth* it ſelf; as ſome Writers have thought.

That the ſaid *Mountains* were not raiſed *ſucceſſively*, and at *ſeveral Times*, being flung up or elevated by *Earthquakes*, ſome at one Time, and ſome at another, as thoſe *Earthquakes* happened. That theſe are ſo far from *raiſing* Mountains, that they *overturn* and *fling down*, ſome of thoſe which were before ſtanding: and *undermine* others, ſinking them into the *Abyſs* underneath *. That of all the *Mountains* of the *whole Globe*, which are very *numerous*, and many of them extremely *large*, and conſequently cannot be ſuppoſed to have been all thus raiſed without the *Notice* of *Mankind*, yet there is not any *authentick Inſtance*, in *all Hiſtory*, of ſo much as *one ſingle Mountain* that was heaved up by an *Earthquake*. That the *new Mountain* in the *Lucrine* Lake, not far from *Pozzuolo* in *Italy*, called *Monte di Cinere*, which is alledged, by the Favourers of *this Opinion*, as an *Inſtance*

* *Confer Part* III. *Sect.* 1. *Conſ.* 12.

ſtance in Behalf of it, was not raiſed thus; the *Relators* of that *Accident*, as well thoſe who were then living, as they who wrote ſince, unanimouſly agreeing that this is not a Mountain conſiſting, as *others* do, regularly of *Strata* *, but a meer *confuſed Heap* of *Stones*, *Cinders*, *Earth*, and *Aſhes*, which were *ſpued* up out of the *Bowels* of the *Earth*, by the Eruption of a *Volcano*, which happened there, in the Year 1538. And though this *Eruption* was preceded by ſeveral *Earthquakes* (the Country all round having been frequently ſhaken for almoſt the Space of two Years before) as thoſe of *Ætna*, *Veſuvius*, and *Hecla* uſualy are, yet *this Hill* was not *elevated* or *heaved up* by any of thoſe *Earthquakes*, but the *Matter*, whereof 'tis compiled, diſcharged out of the *Volcano*, as aforeſaid; in like manner as *Ætna*, *Veſuvius*, and the reſt, fling forth *Stones*, *Cinders*, *&c.* upon any extraordinary *Eruption* of them.

* *Confer pag.* 55: as alſo *Part* 2. *Conſect.* 3. *ſupra.*

That there have not been any *Iſlands* of Note, or conſiderable Extent, *torn* and *caſt off* from the *Continent* by *Earthquakes*, or *ſever'd* from

from it by the boiſterous *Inſults* of the *Sea.* That *Sicily*, *Cyprus*, the *Negropont*, and many more, which have been ſuppoſed by ſome to be only *diſmembred Parcels* of the *Main-Land*, and anciently *parted* from it by one or other of theſe *Means*, yet realy never were ſo: but have been *Iſlands* ever ſince the Time of the *Noetick Deluge.*

Unto this Second Part I ſhall annex,

A *Diſcourſe* concerning the *Trees*, which are commonly called *Subterranean Trees*, or *Foſſil Wood*, and which are found in great Plenty buried amongſt other *Vegetable Bodyes* in *Moſſes**, *Fens*, or *Bogs*, not only in *ſeveral Parts* of *England*, but likewiſe in many *Foreign Countries*; *wherein* I ſhall ſhew, from *Obſervations* made upon the *Places* where theſe *Trees* are digg'd up: upon the *Trees* themſelves: their *Poſition* in the *Earth*, and other *Circumſtances*, that they were lodged thus by the *Deluge*, and have lain *here* ever ſince. That there are found *various Kinds* of theſe *Trees*, and ſeveral of *conſiderable Bulk*, ſo buryed

* *Moſs* is the Name uſed all over the North of *England* inſtead of *Moraſs*, or *Marſh*; of which indeed *Moſs* ſeems to have been a Corruption.

buryed in *Iſlands* where no *Trees* at all do, or will *now* grow; the *Winds* being ſo fierce, and the *Weather* ſo ſevere, as not to ſuffer any thing to proſper or thrive beyond the Height of a Shrub, in any of all thoſe *Iſlands*, unleſs it be *protected* by *Walls*, as in *Gardens*, or other like *Coverture*. That the ſaid *Trees* are in ſome Places found incloſed in the *Stone* of *Quarries* and of *Rocks*: buryed amongſt *Marle*, and *other Kinds* of *Earth*, as well as in this *Peat* or *Moſs-Earth*. That they were *originaly* lodged indifferently amongſt *all Sorts* of *Terreſtrial Matter*, which lay near the *Surface* of the *Earth* *: and that they are at this Day found very ſeldom unleſs in *this Peat-Earth*, is meerly *accidental*; *this Earth* being of a *bituminous* and *mild Nature*; ſo that the *Trees* lay all this while, as it were, *embalm'd* in it, and were by that Means *preſerved* down to *our Times*; whilſt thoſe which chanced to be lodged in *other Earth*, that was more *lax* and *pervious*, *decayed* in tract of Time, and *rotted* at length, and therefore do not now *appear* at

* *Confer Conſect.* 3. ſupra.

at all, when we dig and search into *those Earths*. Or if any thing of them do *appear*, 'tis only the *Ruins*, or some slight *Remains* of them; there being very rarely found any *Trunks* of *Trees*, in these *laxer Earths*, that are *intire*, or tolerably *firm* and *sound*. To conclude, from several of the aforesaid *Circumstances* I shall evince that these *Trees* could never possibly have been *reposed* thus by any other Means than the *Deluge*: neither by *Men*: nor by *Inundations*: nor by *Deterrations* *: nor by violent and impetuous *Winds*: nor by *Earthquakes*; which are the *several Ways* whereby *Learned Men* have thought they were thus *bury'd*.

* *Confer Part* 5. *Consect*. 2.

PART

PART III.

Concerning the Fluids *of the* Globe.

****

SECT. I.

Of the great Abyſs. *Of the* Ocean. *Concerning the Origin of* Springs, *and* Rivers. Of Vapours, *and of* Rain.

HAVING thus done with the more *bulky* and *corpulent Parts* of the *Globe*, the next Place in courſe is due unto *Metalls* and *Minerals*, which are the only remaining Part of the *Terreſtrial Matter* of it not yet treated of. And accordingly I ſhould now

now paſs on to *theſe*; but the preſent Oeconomy and Diſpoſal of ſome of them being wholey owing to the *Motion* and *Paſſage* of *Water* in the interiour Parts of the *Earth*, I have for that Reaſon choſen rather, that I may be as brief as poſſible, and avoid all needleſs Repetitions, to wave *them* for a while, 'till I have firſt offered what I have to ſay about *that*.

The *Water* therefore of the Globe, as well that reſident *in it*, as that which floats *upon it*, is the Subject which I purpoſe here to proſecute. In order whereunto, I ſhall ſub-divide this third Part into *two Sections*; the *former* whereof will comprehend what relates to the preſent and *natural State* of the *Fluids* in and upon the *Earth*: the *other*, what concerns that extraordinary *Change* of this *State* which happen'd at the *Deluge*, and how that *Change* was wrought.

At the Head of the firſt of theſe Sections I prefix a new Set of *Obſervations touching the Fluids of the Terraqueous Globe*: the *Sea*, *Rivers*, and *Springs*: the *Water* of Mines, of of Cole-pits: of Caves, Grotts, and the like Recesses: as also concerning *Vapours*, *Rain*, *Hail*, and *Snow*.

And because this is a *Subject* of that vast *Latitude*, that the Strength of one single Man will scarcely be reckon'd sufficient effectualy to cultivate and carry it on, I have taken in the joint Assistance of *other Hands*, and super-added, to my own, all such *Relations* as I could procure from *Persons* whose Judgment and Fidelity might safely be rely'd upon, about the *Sea*, *Lakes*, *Rivers*, *Springs*, and *Rain*, not only of *this Island*, but many *other Parts* of the *World* besides. Nor do I neglect those which are already extant in the *Published Discourses* of diligent and inquisitive Men.

From all which *Observations*, joyn'd with those made by my self, I prove,

That there is a mighty *Collection of Water* inclosed in the *Bowels of the Earth*, constituting an *huge Orb* in the *interiour* or *central Parts* of it; upon the Surface of which Orb of Water the *Terrestrial Strata* are expanded. That *this* is the same which 1.

which *Moſes* calls the *Great Deep*, or *Abyſs*: the ancient Gentil Writers, *Erebus*, and *Tartarus*.

2. That the *Water* of this Orb *communicates* with *that* of the *Ocean*, by means of certain *Hiatus*'s or *Chaſmes* paſſing betwixt it and the Bottom of the *Ocean*. That they have the ſame *common Center*, around which the *Water* of both of them is compiled and arranged; but in ſuch Manner, that the ordinary *Surface* of this *Orb* is not level with *that* of the *Ocean*, nor at ſo great a *Diſtance* from the *Center* as *that* is, it being for the moſt part reſtrained and *depreſſed* by the *Strata* of Earth lying upon it. But wherever thoſe *Strata* are *broken*, or ſo *lax* and *poroſe* that *Water* can *pervade* them, there the *Water* of the ſaid *Orb* does *aſcend*: fills up all the *Fiſſures* whereinto it can get Admiſſion or Enterance: and *ſaturates* all the Interſtices and Pores of the Earth, Stone, or other Matter, all round the Globe, quite up to the *Level* of the Surface of the *Ocean*.

3. That there is a perpetual and inceſſant *Circulation* of *Water* in the *Atmoſphere*;

Atmoſphere; it *ariſing* from the Globe in Form of *Vapour*, and *falling* down again in *Rain*, *Dew*, *Hail*, and *Snow*. That the *Quantity* of *Water* thus riſing and falling is *equal*; as much *returning* back in *Rain*, *&c.* to the whole *terraqueous Globe*, as was *exhaled* from it in *Vapours*. That tho' the *Quantity* of *Water* thus riſing and falling be certain and conſtant as to the *whole*, yet it varies in the ſeveral *Parts* of the *Globe*; by reaſon that the *Vapours float* in the *Atmoſphere*, *ſailing* in *Clouds* from Place to Place, and are not reſtored down again in a Perpendicular upon the *ſame* preciſe Tract of *Land*, or *Sea*, or *both* together, from which Originaly they *aroſe*, but *any other* indifferently. So that *ſome Regions* receive back *more* in *Rain* than they ſend up in *Vapour*: as, on the contrary, *others* ſend up more in *Vapour* than they *receive* in *Rain*. Nay, the very *ſame Region*, at *one Seaſon*, ſends up *more* in *Vapours* than it receives in *Rain*: and, at *another*, receives *more* in *Rain* than it ſends up in *Vapour*. But the *Exceſſes* of one *Region* and *Seaſon* compenſating the *Defects* of the others, the Quantity *riſing* and *falling* upon the *whole Globe* is *equal*; however *dif-ferent* it may be in the ſeveral *Parts* of it.

4. That the *Rain* which falls upon the *Surface* of the *Earth* partly runs off into *Rivers*, and thence into the *Sea*: and partly *ſinks* down into the *Earth*, inſinuating it ſelf into the Interſtices of the Sand, Gravel, or other Matter of the exteriour or uppermoſt *Strata*. Whence *ſome of it* paſſes on into *Springs*, *Wells*, and into *Grotts*, and *ſtagnates* there, 'till 'tis by Degrees again *exhaled*. *Some of it* glides into the *perpendicu-lar Intervalls* of the ſolid *Strata*; where, if there be no *Outlet* or *Paſ-ſage* to the *Surface*, it *ſtagnates*, as the other; but, if there be ſuch *Outlets*, 'tis by them *refunded forth* together with the ordinary Water of *Springs* and *Rivers*. And the *reſt*, which, by reaſon of the *compactneſs* of the *terreſtrial Matter* underneath, cannot make its Way to *Wells*, the *perpendicular Fiſſures*, or the like *Exits*, only *ſaturates* the upper-moſt *Strata*: and in time *remounts* up

up again in *Vapour* into the *Atmosphere.*

That altho' *Rains* do thus fall into and augment *Springs* and *Rivers*, yet neither *the one* nor *the other* do derive the *Water*, which they ordidinarily refund, from *Rains*; notwithstanding what very many *Learned Men* have believed. 5.

That *Springs* and *Rivers* do not proceed from *Vapours* raised out of the *Sea* by the *Sun*, born thence by *Winds* unto *Mountains*, and there *condensed*; as a modern ingenious *Writer* is of Opinion. 6.

That the abovementioned great subterranean Magazine the *Abyss*, with its Partner the *Ocean*, is the *Standing Fund* and *Promptuary* which supplies *Water* to the Surface of the Earth: as well *Springs* and *Rivers*, as *Vapours* and *Rain*. 7.

That there is a nearly uniform and constant *Fire* or *Heat** dissemi- 8.

nated

* *Heat* and *Fire* differ but in *Degree*: and Heat is Fire, only in lesser Quantity. Fire I shall shew to be a *Fluid* consisting of Parts extremely *small* and *light*, and consequently very *subtile*, *active*, and susceptive of *Motion*. An *Aggregate* of these *Parts* in such *Number* as to be *visible* to the Eye, is what we call *Flame* and *Fire*: a lesser, thinner, and more dispers'd Collection, *Heat* and *Warmth*.

nated throughout the *Body* of the *Earth*, and eſpecialy the *interiour Parts* of it; the Bottoms of the deeper *Mines* being very *ſultry*, and the *Stone* and *Ores* there very ſenſibly *hot*, even in Winter, and the colder Seaſons. That 'tis *this Heat* which *evaporates* and *elevates* the *Water* of the *Abyſs*, buoying it up indifferently on every Side, and towards all Parts of the *Surface* of the *Globe*; pervading not only the *Fiſſures* and *Intervalls* of the *Strata*, but the very *Bodyes* of the *Strata* themſelves, *permeating* the *Interſtices* of the Sand, Earth, or other Matter, whereof they conſiſt: yea even the moſt firm and denſe, *Marble*, and *Sand-ſtone*. For theſe give *Admiſſion* to it, though in leſſer Quantity, and are always found *ſaturated* with it; which is the Reaſon that they are *ſofter*, and cut much more eaſily, when *firſt* taken out of their Beds and Quarries, than afterwards, when they have lain ſome time expoſed to the Air, and that *Humidity* is evaporated.

That this *Vapour* proceeds up *directly* towards the *Surface* of the *Globe*

Globe on *all Sides*, and, as near as poſſible, in *right Lines*, unleſs *impeded* and *diverted* by the Interpoſition of *Strata* of *Marble*, the denſer Sorts of *Stone*, or other like Matter, which is ſo *cloſe* and *compact* that it can admitt it only in *ſmaller Quantity*, and this very ſlowly and leiſurely.

That where the *Vapour* is thus *intercepted* in its Paſſage, and cannot penetrate the *Stratum* diametricaly, *ſome of it* glides along the *lower Surface* of it, permeating the *horizontal Intervall* which is betwixt the ſaid denſe *Stratum* and that which lies underneath it. *The reſt* paſſes the *Interſtices* of the *Maſs* of the *ſubjacent Strata*, whether they be be of *laxer Stone*, of *Sand*, of *Marle*, or the like, with a *Direction* parallel to the *Site* of thoſe *Strata*, 'till it arrives at their *perpendicular Intervalls*.

That this *Water* being thus approach'd to theſe *Intervalls*, in caſe the *Strata*, whereby the aſcending *Vapour* was collected, and *condenſed* into *Water*, as we uſualy ſpeak, in like Manner as by an *Alembick*, happen to be *raiſed* above the *Level* of the Earth's ordinary *Surface*, as thoſe *Strata* are whereof *Mountains* conſiſt, then the *Water*, being likewiſe got *above* the ſaid *Level*, *flows forth* of thoſe *Intervalls* or *Apertures*, and, if there be no Obſtacle without, forms *Brooks* and *Rivers*. But where the *Strata*, which ſo *condenſe* it, are *not higher* than the *mean Surface* of the *Earth*, it *ſtagnates* at the *Apertures*, and only forms *ſtanding-Springs*.

That though the *Supply* from this great *Receptacle* below be *continual*, and nearly the *ſame* at all *Seaſons*, and *alike* to all *Parts* of the *Globe*, yet when it arrives at or near the *Surface* of the *Earth*, where the *Heat* (the *Agent* which *evaporates* and *bears* it up) is *not ſo conſtant* and *uniform* as is that reſident *within* the *Globe*, but is ſubject to *Viciſſitudes* and *Alterations*, being at certain *Seaſons greater* than at *others*: being alſo *greater* in ſome *Climates* and Parts of the *Earth* than in *others*; it hence happens that the *Quantity* of *Water* at the *Surface* of the *Earth*, though ſent up from the

Abyſs

Abyss with an almost constant *Equality*, is *various* and *uncertain*, as is the *Heat* here; at *some Seasons*, and in *some Countryes*, the Surface *abounding*, and being even drown'd with the *Plenty* of it, the *Springs full*, and the *Rivers high*: at *other Seasons*, and in *other Countryes*, both *Springs* and *Rivers exceeding low*, yea sometimes *totaly failing*.

That when the *Heat* in the *exteriour Parts* of the *Earth*, and in the *ambient Air*, is as *intense* as that in the *interiour Parts* of it, *that Water* which passes the Strata *directly*, mounting up in separate Parcels, or in Form of *Vapour*, does *not stop* at the *Surface*; because the Heat *there* is *equal*, both in Quantity, and Power, to *that underneath* which brought it out of the *Abyss*. This *Heat* therefore takes it *here*, and *bears it up*, *part* of it immediately out at the *Surface* of the *Earth*: the *rest*, thro' the Tubes and Vessels of the *Vegetables* which grow thereon, *Herbs*, *Shrubs*, and *Trees*, and along with it a Sort of *vegetative* terrestrial Matter, which it *detaches* from out the uppermost *Stratum* wherein these are planted. *This* it *depoſes* in them, for their *Nutriment*, as it paſſes thro' them*: and iſſuing out at the *Tops* and *Extremities* of them, it marches ſtill on, and is *elevated* up into the *Atmoſphere* to ſuch *Height* that, the *Heat* being there *leſs*, it becomes *condenſed*, unites and combines into ſmall *Maſſes* or *Drops*, and at length *falls down* again in *Rain*, *Dew*, *Hail*, or *Snow**. And for the *Water*, which was *condenſed* at the *Surface* of the *Earth*, and ſent forth collectively into *Standing-Springs* and *Rivers*, *this* alſo ſuſtains a *Diminution* from the *Heat* above; being *evaporated*, more or leſs, in Proportion to the greater or leſſer *Intenſeneſs* of that *Heat*, and the greater or leſſer *Extent* of the *Surface* of the *Water* ſo ſent forth.

* *Vid. Conſ.* 10. *infra.*

* As does likewiſe the *other Part* of it, that aſcended in the *open Air* without; and did not thus paſs thorow *Vegetables.*

That as theſe *Evaporations* are at *ſome Times greater*, according to the *greater Heat* of the *Sun*, ſo wherever they alight again in *Rain*, 'tis as much ſuperiour in *Quantity* to the Rain of *colder Seaſons*, as the *Sun's Power* is *then* ſuperiour to its Power in *thoſe* Seaſons. This is apparent even in theſe *Northern Climes*,

Climes, where the Sun's Power is never very great; our *Rains* in *June*, *July*, and *August*, being much *greater* than those of the *colder* Months: the *Drops larger*, and consequently *heavier:* falling *thicker*, *faster*, and with greater *Force:* striking the Ground, at their Fall, with Violence, and making a mighty Noise: beating down the *Fruit* from the Trees, prostrating and laying *Corn* growing in the Fields: and sometimes so *filling* the *Rivers* as to make them *out-swell* their *Banks*, and lay the neighbouring *Grounds* under *Water*. But much more apparent is it in the more *Southern Regions:* in *Abassinia*, *Nigritia*, *Guinea:* in the *East-Indies:* in *Brasil*, *Paraguay*, and other Countries of *South-America*, to instance in no more. In these the *Sun* shews a much *greater Force:* and their *Rains* (which are *periodical*, happening yearly much about the *same Time*, and lasting *several Months*) fall in such *Quantities* as to be more like *Rivers* descending, than *Showers*. And by *these* are caused those mighty *periodical Inundations* of the *Nile*, the *Niger*,

Niger, the *Rio da Volta*: the *Ganges*: the *Rio de las Amazonas*, the *Rio de la Plata*, and other Rivers of thoſe Countryes; to which *Inundation*, *Egypt*, thro which the *Nile* flows, the *Indies*, and the reſt, owe their *extraordinary Fertility*, and thoſe *mighty Crops* they produce after theſe *Waters* are withdrawn from off their *Fields*; *Rain-water*, as I
* Pag. 50. have already noted*, carrying along with it a ſort of *Terreſtrial Matter* that *fertilizes* the *Land*, as being proper for the Formation of *Vegetables*.

That when the *Heat*, in the *exteriour Parts* of the *Earth*, and in the *ambient Air*, is *leſs* than that in the *interiour*, the *Evaporations* are likewiſe *leſs*. And the *Springs* and *Rivers* thereupon do not only *ceaſe* to
* Confer pag. 140. be *diminiſhed**, in Proportion to the *Relaxation* of the *Heat*, but are much *augmented*; a *great Part* of the *Water*, which aſcends to the *Surface* of the *Earth* in *Vapour*, *ſtopping there*, for *want of Heat* to *mount* it thence up into the *Atmoſphere*, and *ſaturating* the ſuperficial or *uppermoſt Strata* with *Water*; which by Degrees

drains

drains down into *Wells*, *Springs*, and *Rivers*, and ſo makes an *Addition* unto them. And this is the Reaſon that *theſe abound with Water* in the *colder Seaſons* ſo much more than they do in the *hotter*.

That the *Water*, which is thus 9.
diſpenſed to the *Earth* and *Atmoſphere* by the *Great Abyſs*, being carryed down by *Rains* and by *Rivers* into the *Ocean*, which, as hath been ſaid communicates, and ſtands at an *Æquilibrium* with the Abyſs, is by that Means *reſtored back* to that ſubterranean Conſervatory; whence it returns again, in a *continual Circulation*, to the Surface of the Earth, in *Vapours*, and *Springs*.

That the *final Cauſe* of this Di- 10.
ſtribution of *Water*, in ſuch *Quantity*, to *all Parts* of the *Earth* indifferently in *Springs*, *Rivers*, and *Rain*: and of this *perpetual Circulation* and *Motion* of it, is the *Propagation of Bodyes*, *Animals*, *Vegetables*, and *Minerals*, in a *continued Succeſſion*. That for *Animals*, they either feed upon *Vegetables* immediately: or, which comes to the ſame at laſt, upon *other Animals* which have fed upon

upon them. So that *Vegetables* are the firſt and *main Fund:* and, *fit Matter* being ſupplied unto *theſe*, *Proviſion* is thereby made for the *Nouriſhment* of *Animals*; *Vegetables* being no other than ſo many *Machines* ſerving to derive that *Matter* from the *Earth*, to digeſt and prepare it, for *their Uſe*, leiſurely and by little and little, as they can admitt and *diſpoſe* of it, and as it is brought to them by the *Miniſtration* of *this Fluid*. That *Vegetables* being naturaly *fix'd* and *tyed* always to the *ſame Place*, and ſo not able (as *Animals* are) to *ſhift*, and ſeek out after *Matter* proper for their *Nutrition*, 'twas indiſpenſibly neceſſary that it ſhould be *brought to them:* and that there would be ſome *Agent*, thus ready and at hand in *all Places*, to do them that *Office*, and ſo carry on this great and important *Work*. For this *Matter*, being impotent, ſluggiſh, and *inactive*, hath no more Power to *ſtir*, or *move* to *theſe Bodyes*, than *they* themſelves have to *move* unto it. So that it muſt have lain eternaly confined to its *Beds* of *Earth*, and then none of theſe *Bodyes* could ever

ever have been *formed*, were there not *this*, or the like *Agent* to *educe* it *thence*, and bear it unto them. Nor does the *Water*, thus hurry'd about from Place to Place, ſerve only to *carry* the *Matter* unto theſe *Bodyes*, but the *Parts* of it being very *voluble* and *lubricous*, as well as *fine* and *ſmall*, it eaſily *inſinuates* it ſelf into, and placidly *diſtends* the *Tubes* and *Veſſels* of *Vegetables*, and by that Means *introduces* into them the *Matter* it bears along with it, *conveying* it to the *ſeveral Parts* of them ; where *each Part*, by a particular *Mechaniſm* in the *Structure* of it, *detaches* and *aſſumes* thoſe *Particles*, of the *Maſs* ſo conveyed, which are proper for the *Nouriſhment* and *Augmentation* of *that Part*, incorporating *theſe* with it, and letting all the *reſt* paſs on with the *Fluid* ; thoſe *Particles* which are either *ſuperfluous*, and *more* than the Parts of the Plant can admitt and manage at one Time: or that are not *ſuitable* and *proper* for the *Nouriſhment* of any of the *Parts* of a *Plant* of *that Kind*, going out at the *Extremities* of it along with the *Water* *. And this

* *Confer Pag.* 140.

this *latter Office* it does likewiſe to *Animals*; *Water*, and other *Fluids*, ſerving to *convey* the *Matter*, whereby they are *nouriſhed*, from their *Stomachs* and *Guts*, thro' the *Lacteals* and finer *Veſſels*, to the *ſeveral Parts* of their *Bodyes*. But the *Formation* of *Animals* and *Vegetables*, being a Thing ſomewhat foreign to my *preſent Purpoſe*, I ſhall adjourn the fuller *Conſideration* of it to another *Occaſion*. How far *Water* is con-concern'd in the *Formation* of *Minerals*, will appear more at large in the *ſucceeding Part* of *this Work*.

11. That 'tis this *Vapour*, or *ſubtile Fluid*, that, aſcending thus inceſſantly out of the *Abyſs*, and *pervading* the *Strata* of Gravel, Sand, Earth, Stone, and the reſt, by Degrees *rots* and *decays* the *Bones*, *Shells*, *Teeth*, and other Parts of *Animals*: as alſo the *Trees*, and other *Vegetables*, which were lodg'd in thoſe *Strata* at the *Deluge* *; this *Fluid*, by its continual *Attrition*, as it paſſes ſucceſſively by them, *fretting* the ſaid *Bodyes*, by little and little *wearing* off and *diſſipating* their *conſtituent Corpuſcles*, and at length quite

* *Part* 2. *Conſect*. 3.

quite *diſſolving* and *deſtroying* their *Texture.* That yet it hath not *this Effect* indifferently upon all of them; thoſe which happened to be repoſed in the *firmer* and *compacter Strata,* e. g. of *Marble,* the *cloſer Kinds* of *Sand-Stone, Chalk,* and the like, being thereby *protected* in great Meaſure from its *Attacks.* For it paſſes through theſe only in *leſſer Quantity,* and that *ſlowly,* and with *Difficulty**. So that its *Motion here* being more *feeble* and *languid,* the *Shells* and other *Bodyes* enclos'd *in theſe* are uſualy found very *firm* and *intire,* many of them retaining even their *natural Colours* to this Day, though they have lain thus above four thouſand *Years:* and may doubtleſs endure much *longer,* even as long as thoſe *Strata,* to which they owe their *Preſervation,* ſhall themſelves *endure,* and continue *intire* and *undiſturb'd.* Whilſt *thoſe* which were lodg'd in *Loam, Sand, Gravel,* and the like more *looſe* and *pervious* Matter, are ſo *rotted* and *decayed,* that they are now not at all, or very difficultly, diſtinguiſhable from the *Loam,* or other *Mat-*

ter

* *Vid. Conſect.* 8. *ſupra.*

ter in which they lye. Not but that there are ſometimes found, even in theſe *laxer Strata*, *Shells*, *Teeth*, and other *Bodyes* that are ſtill tolerably *firm*: and that have eſcaped pretty ſafe. But *theſe* are only ſuch as are of a more than ordinary *robuſt* and *durable Conſtitution*, whereby they were enabled the better to withſtand therepeated *Aſſaults* of the permeating *Fluid*, and to maintain their *Integrity*, whilſt the other *tenderer* Kinds *periſh'd* and were *deſtroyed*.

That this *ſubtile Fluid* exerts the *ſame Power* upon the *Surface* of the *Earth*, that it does in the *Bowels* of it. For as it is inſtrumental to the *Formation* of Bodyes *here* *, ſo is it likewiſe (by a *different Operation*, which I have not Room to deſcribe in this Place) of the *Deſtruction* of them. And that *Corroſion* and *Diſſolution* of *Bodyes*, even the moſt *ſolid* and *durable*, which is vulgarly aſcribed to the *Air*, is cauſed meerly by the *Action of this Matter* upon them; the *Air* being ſo far from *injuring* and *preying* upon the *Bodyes* it *environs*, that it contributes to their

* *Confer Conſ.* 10. *ſupra.*

their *Security* and *Preſervation*, by *impeding* and *obſtructing* the *Action* of this *Matter*. Were it not indeed for the *Interpoſition* of the *Air*, they could never be able to make ſo *long* and *vigorous Reſiſtance* as now they do.

That this *Subterranean Heat* or *Fire*, which thus *elevates* the *Water* out of the *Abyſs*, being in any Part of the Earth *ſtop'd*, and ſo *diverted* from its *ordinary Courſe*, by ſome accidental *Glut* or *Obſtruction* in the *Pores* or *Paſſages* through which it uſed to *aſcend* to the *Surface*: and being by that Means preternaturaly *aſſembled*, in *greater Quantity* than uſual, into *one Place*, it cauſes a great *Rarifaction* and *Intumeſcence* of the *Water* of the *Abyſs*, putting it into very great *Commotions* and *Diſorders*. And at the *ſame Time* making the like *Effort* upon the *Earth*, which is expanded upon the Face of the *Abyſs*, it occaſions that *Agitation* and *Concuſſion* of it, which we call an *Earthquake*. 12.

That this *Effort* is in ſome Earthquakes ſo *vehement* that it *ſplits* and *tears* the *Earth*, making *Cracks* or

Chasmes in it some *Miles* in Length; which *open* at the *Instants* of the *Shocks*, and *close* again in the *Intervalls* betwixt them. Nay, 'tis sometimes so extremely *violent*, that it plainly *forces* the superincumbent *Strata*: *breaks* them all throughout, and thereby perfectly *undermines* and *ruins* the *Foundations* of them. So that, *these failing*, the *whole Tract*, as soon as ever the *Shock* is over, *sinks* down to rights into the *Abyss* underneath, and is *swallowed* up by it; the *Water* thereof immediately *rising* up, and forming a *Lake* in the Place where the said *Tract* before was. That several considerable *Tracts* of Land, and some with *Cityes* and *Towns* standing upon them: as also *whole Mountains*, many of them very *large*, and of a *great Height*, have been thus totaly *swallowed up*.

That this *Effort* being made in *all Directions* indifferently, upwards, downwards, and on every Side; the *Fire* dilating and expanding on *all Hands*, and endeavouring, proportionably to the *Quantity* and *Strength* of it, to get Room, and make

make its Way through *all Obſtacles*, falls as foul upon the *Water* of the *Abyſs* beneath, as upon the *Earth* above; *forcing it forth* which Way ſoever it can find *Vent* or *Paſſage*: as well through its *ordinary Exits*, *Wells*, *Springs*, and the *Outlets* of *Rivers*, as through the *Chaſmes* then newly opened: through the *Camini* or *Spiracles* of *Ætna*, or other near *Volcanoes*: and thoſe *Hiatus*'s at the *Bottom* of the *Sea* *, whereby the *Abyſs* below *opens* into it, and *communicates* with it.

* *Vid. Conſect.* 2. *ſupra.*

That as the *Water reſident* in the *Abyſs* is, in *all Parts* of it, ſtored with a conſiderable *Quantity* of *Heat*: and more eſpecialy in *thoſe* where theſe extraordinary *Aggregations* of this *Fire* happen; ſo likewiſe is the *Water* which is thus *forced out* of it. Inſomuch that, when *thrown forth*, and *mix'd* with the Waters of *Wells*, of *Springs*, of *Rivers*, and the *Sea*, it renders them very ſenſibly *hot*.

That it is uſualy *expelled* forth in *vaſt Quantities*. and with ſo *great Impetuoſity*, that it hath been ſeen to *ſpout up* out of *deep Wells*, and *fly forth*, at the *Tops* of them, upon the Face of the Ground. With like *Rapidity* it comes out of the *Sources* of *Rivers*; *filling* them ſo of a *ſudden* as to make them *run over* their *Banks*, and *overflow* the neighbouring *Territories*, without ſo much as one Drop of *Rain* falling into them, or any other concurrent *Water* to raiſe and augment them. That it *ſpues* out of the *Chaſmes*, opened by the *Earthquake*, in great *Abundance*; mounting up, in mighty *Streams*, to an incredible *Height* in the Air: and this oftentimes at many *Miles* Diſtance from any *Sea*. That it likewiſe *flies forth* of the *Volcanoes* in *vaſt Floods*, and with wonderful *Violence*. That 'tis *forced* through the *Hiatus*'s at the *Bottom* of the *Sea* with ſuch *Vehemence*, that it puts the *Sea* into the moſt horrible *Diſorder* and *Perturbation* imaginable, even when there is not the leaſt Breath of *Wind* ſtirring, but all, till then, *calm* and *ſtill*; making it rage and *roar* with a moſt hideous and amazing *Noiſe*: raiſing its *Surface* into prodigious *Waves*, and toſſing and rowling them about in a very a very ſtrange and furious Manner: overſetting *Ships* in the Harbours, and *ſinking* them to the Bottom; with many other like *Outrages*. That 'tis refunded out of theſe *Hiatus*'s in ſuch *Quantity* alſo that it makes a vaſt *Addition* to the *Water* of the Sea; *raiſing* it many Fathoms *higher* than ever it *flows* in the higheſt Tides, ſo as to pour it forth far beyond its uſual *Bounds*, and make it *overwhelm* the adjacent *Country*; by this Means ruining and deſtroying *Towns* and *Cityes*: drowning both *Men* and *Cattle*: breaking the Cables of *Ships*, driving them from their *Anchors*, ſometimes bearing them along with the *Inundation* ſeveral *Miles* up into the *Country*, and there running them aground: ſtranding *Whales* likewiſe, and other great *Fiſhes*, and leaving them, at its *Return*, upon *dry Land*.

That theſe *Phænomena* are not *new*, or peculiar to the *Earthquakes* which have happen'd in *our Times*, but have been obſerv'd in *all Ages*: and particularly theſe exorbitant *Commotions* of the *Water* of the

Globe. This we may learn abundantly from the *Hiſtoryes* of *former Times:* and 'twas for *this Reaſon* that many of the *Ancients* concluded, rightly enough, that *they* were cauſed by the *Impulſes* and *Fluctuation* of *Water* in the *Bowels* of the *Earth.* And therefore they very frequently called *Neptune* Σεισίχθων, as alſo Ἐνοσίχθων, Ἐνοσίγαιος, and Τιναϰτορογαιης; by all which *Epithets* they denoted his *Power* of *Shaking* the *Earth.* They ſuppoſed that he *preſided* over *all Water* whatever, as well that *within* the *Earth*, as the *Sea*, and the reſt *upon it:* and that the *Earth* was *ſupported* by *Water*, its *Foundations* being laid thereon. Upon which Account it was they beſtow'd upon him the Cognomen Γαιηοχος, or *Supporter of the Earth*, and that of Θεμελιοῦχος, or *The Suſtainer of its Foundations.* They likewiſe believed that he, having a full *Sway* and *Command* over the *Water*, had Power to *ſtill* and *compoſe* it, as well as to *move* and *diſturb* it, and the *Earth* by means of it. And therefore they alſo gave him the Name of Ἀσφάλιος, or *The Eſtabliſher*; under

under which Name ſeveral *Temples* were conſecrated to him, and *Sacrifices* offer'd, whenever an *Earthquake* happened, to pacify and appeaſe him; requeſting that he would allay the *Commotions* of the *Water*, ſecure the *Foundations* of the *Earth*, and put an End to the *Earthquake*.

That the *Fire*, it ſelf, which, being thus *aſſembled* and *pent up*, is the *Cauſe* of all theſe *Perturbations*, makes its *own Way* alſo forth, by what *Paſſages* ſoever it can get vent: through the *Spiracles* of the next *Volcano* *: through the *Cracks* and Openings of the *Earth* above-mentioned: through the *Apertures* of *Springs*, eſpecialy thoſe of the *Thermæ* *: or any other *Way* that it can either find or make. And being thus *diſcharged*, the *Earthquake* ceaſeth, till the *Cauſe* returns again, and a freſh Collection of this *Fire* committs the ſame *Outrages* as before.

* *Confer Conſ.* 13. *infra.*

* *Vid. Conſect.* 14. *infra.*

That there is ſometimes in *Commotion* a Portion of the *Abyſs* of that *vaſt Extent*, as to ſhake the *Earth* incumbent upon it for ſo very *large* a *Part* of the *Globe* together, that the *Shock* is felt the *ſame Minute* preciſely in *Countryes* that are many hundreds of *Miles* diſtant from each other: and this even though they happen to be *parted* by the *Sea* lying betwixt them. Nay, there want not Inſtances of ſuch an univerſal *Concuſſion* of the *whole Globe* as muſt needs imply an Agitation of the *whole Abyſs*.

That though the *Abyſs* be liable to theſe *Commotions* in *all Parts* of it, and therefore no *Country* can be wholey exempted from the *Effects* of them, yet theſe *Effects* are no where very remarkable, nor are there uſualy any great *Damages* done by *Earthquakes*, except only in thoſe *Countryes* which are *mountainous*, and conſequently *ſtoney*, and *cavernous* underneath *: and eſpecialy where the *Diſpoſition* of the *Strata* is ſuch that thoſe *Caverns* open into the *Abyſs*, and ſo freely admitt and entertain the *Fire*, which, aſſembling therein, is the *Cauſe* of the *Shock*; it naturaly ſteering its Courſe *that Way* where it finds the readyeſt *Reception*, which is towards theſe *Caverns*; this being indeed much

* *Vid.* Part 2. Conſect. 8.

much the Caſe of *Damps* in *Mines*, whereof more by and by. Beſides that thoſe Parts of the Earth which abound with *Strata* of *Stone*, or *Marble*, making the ſtrongeſt *Oppoſition* to this *Effort*, are the moſt furiouſly *ſhatter'd*, and ſuffer much more by it than thoſe which conſiſt of *Gravel*, *Sand*, and the like *laxer Matter*, which more eaſily *give way*, and make not ſo great *Reſiſtance*; an *Event* obſervable not only in this, but all other *Exploſions* whatever. But, above all, thoſe *Countryes*, which yield great Store of *Sulphur* and *Nitre*, are by far the moſt *injured* and *incommoded* by *Earthquakes*; theſe *Minerals* conſtituting in the *Earth* a kind of *Natural Gunpowder*, which, taking *fire*, upon this *aſſembly* and *approach* of it, occaſions that murmuring *Noiſe*, that *ſubterranean Thunder*, if I may ſo ſpeak, which is heard rumbling in the *Bowels* of the *Earth* during *Earthquakes*, and, by the Aſſiſtance of its *exploſive Power*, renders the *Shock* much *greater*, ſo as ſometimes to make miſerable *Havock* and *Deſtruction*. And 'tis for this Reaſon that *Italy*, *Sicily*, *Anatolia*,

Anatolia, and some Parts of *Greece*, have been *so long* and *so often* alarmed and harrassed by *Earthquakes*; these Countryes being all *Mountainous* and *Cavernous*, abounding with *Stone* and *Marble*, and affording *Sulphur* and *Nitre* in great *Plenty*. But for a more particular *History* of the several *Phænomena* which precede, which accompany, and which follow after *Earthquakes*: for the *Causes* of those *Phænomena*: and for a more exact *Account* of the *Origine*, and the *Oeconomy* of this *subterranean Fire*, I must beg the Reader's Patience 'till the *larger Work* be made publick.

13. That *Ætna*, *Vesuvius*, *Hecla*, and the other *Volcanoes*, are only so many *Spiracles*, serving for the *Discharge* of this *subterranean Fire*, when 'tis thus preternaturaly *assembled*. That where there happens to be such a *Structure* and *Conformation* of the *interiour Parts* of the *Earth* that the *Fire* may pass, *freely* and without *Impediment*, from the *Caverns*, wherein it *assembles*, unto these *Spiracles*, it then readily and easily *gets out*, from Time to Time,

Time, without *ſhaking* or *diſturbing* the *Earth*. But where ſuch *Communication* is wanting, or the *Paſſages* not ſufficiently *large* and *open*, ſo that it cannot come at the ſaid Spiracles without firſt *forcing* and *removing* all *Obſtacles*, it *heaves* up and *ſhocks* the Earth, with greater or leſſer *Impetuoſity*, according as the *Quantity* of the *Fire* thus aſſembled is greater or leſs, till it hath made its Way to the Mouth of the *Volcano*; where it ruſheth forth, ſometimes in mighty *Flames*, with great *Velocity*, and a terrible bellowing *Noiſe*. That therefore there are ſcarcely any *Countryes*, that are much *annoyed* with *Earthquakes*, that have not one of theſe *Firey Vents*. And *theſe* are conſtantly all in *Flames* whenever any *Earthquake* happens; they *diſgorging* that *Fire*, which, whilſt *underneath*, was the *Cauſe* of the *Diſaſter*. Indeed, were it not for theſe *Diverticula*, whereby it thus gains an *Exit*, 'twould *rage* in the *Bowels* of the *Earth* much more *furiouſly*, and make greater *Havock* than now it doth. So that though *thoſe Countryes*,

tryes, where there are ſuch *Volcanoes*, are uſualy, more or leſs, troubled with *Earthquakes* ; yet, were theſe *Volcanoes* wanting, they would be much more *annoy'd* with *them* than now they are ; yea in all probability to that *Degree*, as to render the Earth, for a vaſt Space round them, perfectly *uninhabitable*. In one Word, ſo beneficial are *theſe* to the *Territories* where they are, that there do not want *Inſtances* of *ſome* which have been *reſcu'd* and wholey *deliver'd* from *Earthquakes* by the breaking forth of a new *Volcano* there ; this continualy *diſcharging* that *Matter*, which, being till then *barricaded* up, and *impriſoned* in the *Bowels* of the *Earth*, was the Occaſion of very great and frequent *Calamityes*. That moſt of theſe *Spiracles* perpetualy and at *all Seaſons* ſend forth *Fire*, more or leſs: and though it be ſometimes ſo little that the Eye cannot diſcern it, yet even then, by a nearer Approach of the Body, may be diſcover'd a copious and very ſenſible *Heat* continualy iſſuing out.

That

That the *Thermæ*, *Natural Baths*, or *Hot-Springs*, do not owe their *Heat* to any *Colluctation* or *Effervescence* of the *Minerals* in them, as some *Naturalists* have believ'd: but to the beforemention'd *Subterranean Heat* or Fire. That these *Baths* continualy *emitt* a *manifest* and very *sensible Heat*: nay some of them have been observed at some Times to send forth an *actual* and *visible Flame*. That not only *these*, but all other *Springs* whatever, have in them some Degree of *Heat**, (none of them ever *Freezing*, no not in the longest and severest *Frosts*) but more especialy those which arise where there is such a *Site* and *Disposition* of the *Strata* within the *Earth* as gives free, and easy *Admission* to this *Heat*, and favours its *Ascent* to the *Surface*; where, *perspiring forth* at the same *Outlets* with the *Water* of the *Spring*, it by that Means *heats* it, more or less, as it chanceth to be dispensed forth in greater or lesser *Quantity*. That as the *Heat* of *all Springs* is owing to this *subterraneous Fire*, so wherever there are any *extraordinary Dischar-* 14.

* It is indeed by this very *Heat* that their *Water* is born unto them from out the *Abyss*. *Vid.* Conf. 8. *supra.*

ges of *this Fire*, there alſo are the neighbouring Springs *hotter* than *ordinary*; witneſs the many *Hot-Springs* near *Ætna*, *Veſuvius*, *Hecla*, and all other *Volcanoes*. That the *Heat* of the *Thermæ* is not *conſtant*, and always *alike*; the *ſame Spring* ſuffering at ſome Times a very manifeſt *Failure* and *Remiſſion* of its *Heat*: at others as manifeſt an *Addition* and *Encreaſe* of it; yea ſometimes to that Exceſs as to make it *boil* and *bubble* with *extream Heat*, like *Water* when boiling over a *common Fire*. That particularly during *Earthquakes*, and Eruptions of *Volcanoes* *, when there is a more copious *Acceſſion* of this *ſubterraneous Fire*, the *Thermæ* all thereabouts become much *hotter* than before; yielding alſo a far *greater Supply* of *Water* than they were wont to do: and a murmuring *Noiſe* is uſualy heard, below them, in the *Bowels* of the *Earth*. All which is occaſioned meerly by the then rapid *Motion*, and *Aſcent* of the *Fire*, in greater *Plenty* than before, to the *Apertures* of theſe *Springs*.

* *Vid.* *Conſ.* 12. and 13. *ſupra.*

I have

I have now finish'd the Account of *this Section:* and was just going to take off my Hand here. But recollecting that, in the foregoing Part of this Work*, I promised some further *Proofs* of *Contrivance* in the *Structure* of the *Globe* we dwell upon: and such too as may satisfy any fair and unbyass'd Spectator that the *Frameing* and *Composition* of it out of the *Materials* of the *former Earth* was a *Work* of *Counsel* and *Sagacity:* a *Work* apparently above the highest Reaches of *Chance*, or the Powers of *Nature*; and *this* being a proper Place wherein to produce those *Proofs*, I shall give such Hints of them as the Brevity I am tyed up to will permitt me, and then conclude.

* *Pag.*

I am indeed well aware that the Author of the *Theory of the Earth** differs very much from me in Opinion as to this Matter. He will not allow that there are any such Signs of Art and Skill in the Make of the present Globe as are here mention'd: or that there was so great Care, and such exact Measures taken in the re-fitting of it up again at

* *Lib.* 1. *Cap.* 9. to 12.

at the Deluge. He reckons it no other than an huge diſorderly Pile of *Ruines and Rubbiſh*: and is very unwilling to believe that it was the Product of any Reaſoning or Deſigning Agent. The Chanel of the Ocean appears to him *the moſt ghaſtly Thing in Nature*, and he *cannot at all admire its Beauty or Elegancy*: *for 'tis*, in his judgment, *as deformed and irregular as it is great*. And for the Caverns of the Earth, the Fiſſures and Breaches of the *Strata*, he cannot fancy that they were formed *by any Work of Nature*, *nor by any immediate Action of God*, *ſeeing there is neither Uſe*, that he can diſcover, *nor Beauty in this Kind of Conſtruction*. Then for the Mountains, *theſe*, he ſays, *are placed in no Order one with another*, *that can either reſpect Uſe or Beauty*, *and do not conſiſt of any Proportion of Parts that is referable to any Deſign*, *or that hath the leaſt Footſteps of Art or Counſel*. In fine, he thinks there are ſeveral Things in the Terraqueous Globe that are *rude and unſeemly*: and many that are *ſuperfluous*. He looks upon it as *incommodious*, and

as

as a broken and confused Heap of Bodyes, placed in no Order to one another, nor with any Correspondency or Regularity of Parts: and it seems, to him, nothing better than a *rude Lump*, and a *little dirty Planet.* I have given his Opinion in his own Words, though I have upon all like Occasions taken a shorter Course, and contented my self with giving only the Sense of Others; but this I have done, here, least any Man should suspect that I mistake the Author's Sentiments, or do not represent them fairly.

Now though it were realy so, that there were some such Eye-Sores in our *Earth* as are here suggested: and that we could not presently find out all the Gayetyes and *Embelishments* that we might seek for in it, the Matter would not be great: and we might very well be contented to take it as we find it. But after all, the Thing is quite otherwise, and there are none of all these wanting: nor any such *Deformityes* as are here imagined; but, on the contrary, so very many real Graces and *Beautyes*, that 'tis no easy Thing to overlook them all. Even this very Variety of *Sea* and *Land*, of *Hill* and *Dale*, which is here reputed so inelegant and unbecoming, is indeed extreamly charming and agreeable. Nor do I offer this as any private Fancy of my own, but as the *common Sense* of *Mankind*, who are the true and proper Judges in the Case; both the *Antients* and *Moderns*, giving their Suffrages unanimously herein. Even the very *Heathens* themselves, have esteemed this *Variety* not only *ornamental* to the Earth, but a Proof of the *Wisdom* of the *Creator* of it, and *alledged* it as such; whereof more in due Place.

And, as I cannot admitt that there is any thing *unhandsome* or *irregular*: so much less can I grant that there is any thing *incommodious* and *Artless*, or *useless* and *superfluous*, in the *Globe*. Were I at full Liberty to do it here, 'twould be no hard thing to make appear that there are no real Grounds for any such *Charge*. For how easy were it, by taking a minute and distinct Survey of the *Globe*, and of the very many and various

various *Limbs* and *Parts* of it, to ſhew that all theſe are order'd and digeſted with infinite *Exactneſs* and *Artifice*; each in ſuch Manner as may beſt ſerve to its *own* proper *End*, and to the *Uſe* of the *whole*? How eaſy were it to ſhew, that the *Rocks*, the *Mountains*, and the *Caverns*, againſt which theſe *Exceptions* are made, are of indiſpenſible *Uſe* and *Neceſſity*, as well to the *Earth* as to *Man* and other *Animals*, and even to all the reſt of its *Productions*? that there are no ſuch *Blemiſhes*, no *Defects*: nothing that might have been *alter'd* for the *better*: nothing *ſuperfluous*: nothing *uſeleſs*, in all the whole Compoſition? and ſo finaly trace out the numerous *Footſteps* and *Marks* of the *Preſence* and *Interpoſition* of a moſt wiſe and intelligent *Architect* throughout all this realy wonderfull *Fabrick*? But I muſt reſerve *this* for a fitter Opportunity, and content my ſelf for the preſent with only giving ſome brief Intimations of it in the *following Propoſitions*. Namely,

That 'twas abſolutely neceſſary for the *well being* both of the *Earth* it ſelf, and of all *Terreſtrial Bodyes*, that ſome of the *Strata* ſhould *conſolidate*, as they did, immediately after the Subſidence of their Matter at the *Deluge:* that *theſe* ſhould afterwards be *broken* in certain Places: and laſtly, that they ſhould be *diſlocated*, ſome of them *elevated*, and others *depreſſed*.

That had not the *Strata* of Stone and Marble become *ſolid**, but the Sand, or other Matter whereof they conſiſt, continued *lax* and *incoherent*, and they conſequently been as *pervious* as thoſe of Marle, Gravel, and the like, the *Water* which riſes out of the *Abyſs*, for the Supply of *Springs* and *Rivers*, would not have *ſtop'd* at the *Surface* of the *Earth*, but *march'd* directly, and without Impediment, up into the *Atmoſphere*, in all Parts of the Globe, wherever there was *Heat* enough in the Air to continue its *Aſcent*, and buoy it up. So that there then muſt needs have been an univerſal *failure* and *want* of *Springs* and *Rivers* all the *Summer-Seaſon*, in the *colder Climes:* and *all the Year* round in the *hotter*, and thoſe that are near

* *As Pt.* 2. *Conſect.* 4.

near the *Æquator*, where there is much the greateſt *need* of both the one and the other; and this meerly for want of the Interpoſition of ſuch *denſe* and *ſolid Strata*, to *arreſt* the aſcending *Vapour*, to ſtop it at the *Surface* of the *Earth*: and to *collect* and *condenſe* it there.

That though the *Strata* had become ſolid, ſo as to have condenſed the riſing Vapour, yet if they had not been *broken* alſo*, the Water muſt have lain eternaly *underneath* thoſe *Strata*, without ever *coming forth*. And conſequently there then could have been neither *Springs* nor *Rivers* for a very conſiderable Part, or indeed, almoſt the whole *Earth*; the *Water*, which ſupplyes theſe, proceeding out at thoſe *Breaches**. This *Water* therefore would have been wholey *intercepted*, all lock'd up within the Earth, and its Egreſs utterly debarr'd, had the *Strata* of Stone and Marble remained *continuous*, and without ſuch *Fiſſures* and *Interruptions*. That theſe *Fiſſures* have a ſtill *further Uſe*, and ſerve for *Receptacles* of *Metalls*, and of ſeveral Sorts of *Minerals*;

* As *Pt.* 2. *Conſect.* 6.

* *Conſ.* 8. *ſupra.*

which are *arrested* by the *Water* in its *Passage* thither thro' the *Strata* wherein the single *Corpuscles* of those *Metalls* and *Minerals* were lodg'd*: and *borne* along with it into these *Fissures*; where, being by this Means *collected*, they are kept in Store for the *Use* of *Mankind*.

* Part 4. Consect. 5.

That though there had been both *solid Strata* to have *condens'd* the ascending Vapour: and *those* so *broken* too as to have given free *Vent* and *Issue* to the *Water* so condens'd; yet had not the said *Strata* been *dislocated* likewise*, some of them *elevated*, and others *depress'd*, there would have been no *Cavity* or *Chanel* to give Reception to the Water of the *Sea*: no *Rocks*, *Mountains*, or other Inequalities in the *Globe*. And without *these*, the *Water*, which now *arises* out of it, must have all *stagnated* at the *Surface*, and could never possibly have been *refunded forth* upon the *Earth*: nor would there have been any *Rivers*, or *running Streams*, upon the Face of the *whole Globe*, had not the *Strata* been thus *raised up*, and the *Hills exalted* above the neighbouring *Valleys*

* Part 2. Conf. 6, 7, 8.

leys and *Plains*; whereby the *Heads* and *Sources* of *Rivers*, which are in those *Hills*, were also *borne up* above the ordinary *Level* of the *Earth*, so that they may *flow* upon a *Descent*, or an *inclining Plane*, without which they could not *flow* at all *.

That this *Affair* was not transacted unadvisedly, casualy, or at random; but with due *Conduct*, and just Measures. That the *Quantity* of Matter consolidated: the *Number*, *Capacity*, and *Distances* of the *Fissures*: the *Situation*, *Magnitude*, and *Number* of the *Hills*, for the condensing, and discharging forth the *Water*: and, in a Word, all other Things were so order'd that they might *best* conduce to the *End* whereunto they were *design'd* and *ordain'd*. There was such *Provision* made, that a *Country* should not want so many *Springs* and *Rivers* as were *convenient* and *requisite* for it: nor, on the other Hand, be *overrun* with them, and afford little or nothing else; but that there should be a *Supply* every where ready, *suitable* to the *Necessities* and *Expences* of each *Climate* and *Region* of the

* *Confer Consect.* 8. supra.

Globe. For Example, thoſe *Countryes* which lye in the *Torrid Zone*, and under or near the *Line*, where the *Heat* is very *great*, are furniſh'd with *Mountains* anſwerable: *Mountains* which both for *Bigneſs* and *Number* ſurpaſs thoſe of *colder Countryes* as much as the *Heat* there ſurpaſſes that of *thoſe Countryes.* Witneſs the *Andes*, that prodigious Chain of Mountains in *South America: Atlas* in *Africa: Taurus* in *Aſia:* the *Alpes* and *Pyrenees* of *Europe*, to mention no more. By *theſe* is *collected* and *diſpenſed* forth a Quantity of *Water* proportionable to the *Heat* of *thoſe Parts.* So that although, by reaſon of the *Exceſs* of this *Heat* there, the *Evaporations* from the *Springs* and *Rivers* are *very great*; yet they, being, by theſe *larger Supplyes*, continualy *ſtock'd* with an *Exceſs* of Water *as great*, yield a Maſs of it for the Uſe of *Mankind*, the *Inhabitants* of *thoſe Parts*, of the other *Animals*, and of *Vegetables*, not much, if at all, inferiour to the Springs and Rivers of *colder Climates.* That, beſides this, the Waters thus evaporated and mounted ed up into the Air, *thicken* and *cool* it, and, by their *Interposition* betwixt the *Earth*, and the *Sun*, ſkreen and fence off the *ardent Heat* of it, which would be otherwiſe *unſupportable:* and are at laſt returned down again in copious and fruitful *Showers* to the *ſcorched Earth*; which, were it not for this remarkably *Providential Contrivance* of Things, would have been *there* perfectly *uninhabitable:* laboured under an eternal *Drought:* and have been continualy *parched* and *burnt.*

To this former Section I ſhall add, by way of Appendix,

A Diſſertation concerning the *Flux and Reflux of the Sea:* and its other Natural *Motions*; with an Account of the *Cauſe* of thoſe *Motions:* as alſo of the *End* and *Uſe* of them: and an Enquiry touching the *Cauſe* of the *Ebbing* and *Flowing*, and ſome other uncommon *Phœnomena* of certain *Springs*. 1.

A Diſcourſe concerning the *Saltneſs* of the *Sea*. 2.

A Diſcourſe concerning *Wind:* the *Origin*, and Uſe of it in the Natural World. 3.

SECT.

SECT. II.

Of the Univerſality *of the* Deluge. *Of the* Water *which effected. it. Together with ſome further* Particulars *concerning it.*

IN the precedent Section I conſider the *preſent* and *natural State* of the *Fluids* of the *Globe*. I ranſack the ſeveral Caverns of the *Earth*: and ſearch into the Storehouſes of *Water*; and this principaly in order to find out where that mighty *Maſs* of *Water*, which *overflow'd* the *whole Earth* in the Days of *Noah*, is *now* beſtow'd and conceal'd: as alſo which Way 'tis at *this Time* uſeful to the *Earth* and its *Productions*, and ſerviceable to the *preſent Purpoſes* of Almighty *Providence*.

Such a *Deluge* as that which *Moſes* repreſents, whereby *All the high Hills*

*Hills that were under the whole Heaven were cover'd**, would require a portentous Quantity of Water: and Men of Curiosity, in all Ages, have been very much to seek what was become of it, or where it could every find a *Reservatory* capable of containing it. 'Tis true there have been several who have gone about to inform them, and set them to rights in this Matter; but for want of that Knowledge of the *present System* of *Nature*: and that Insight into the *Structure* and *Constitution* of the *Terraqueous Globe*, which was necessary for such an Undertaking, they have not given the Satisfaction that was expected. So far from it that the *greatest Part* of these, seeing no where Water enough to effect a *General Deluge*, were forc'd at last to mince the Matter, and make only a *Partial* one of it; restraining it to one *single Country*: to *Asia*, or some lesser Portion of Land; than which, nothing can be more contrary to the *Mosaick* Narrative.

* *Gen.* vii. 19.

For *the rest*, they had Recourse to *Shifts* which were not much better: and rather evaded than solved the

Difficulty;

Difficulty; ſome of them imagining that a Quantity of Water, ſufficient to make ſuch a *Deluge*, was *created* upon that Occaſion: and, when the Buſineſs was done, all diſbanded again and *annihilated*. Others ſuppoſed a *Converſion* of the *Air* and *Atmoſphere* into *Water*, to ſerve the turn. Many of them were for fetching down I know not what *ſupercæleſtial Waters* for the Purpoſe. Others conluded that the *Deluge* roſe only *fifteen Cubits* above the *Level* of the Earth's *ordinary Surface*, covering the *Valleys* and *Plains*, but not the *Mountains*; all equaly wide of *Truth*, and of the Mind of the *Sacred Writer*.

One of the laſt *Undertakers* of all, ſeeing this, began to think the *Cauſe* deſperate: and therefore, in Effect, gives it up. For, conſidering how unſucceſsful the Attempts of thoſe who were gone before him had proved: and having himſelf * alſo *employ'd his laſt and utmoſt Endeavours to find out Waters for the vulgar Deluge*: having muſter'd up all the Forces he could think of, and all too little: *the Clouds above, and the*

* *Theory of the Earth*, l. 1. c. 2.

the Deeps below, and in the Bowels of the Earth; *and these*, says he, *are all the Stores we have for Water, and Moses directs us to no other for the Causes of the Deluge*; he prepares for a Surrender, asserting, from a mistaken and defective Computation, that *all these* will not come up to near the Quantity requisite: and that *in any known Parts of the Universe, to find Water sufficient for this Effect, as it is generaly explained and understood, is*, he thinks, *impossible*: that is, sufficient to cause *a Deluge*, to use his own Words, *overflowing the whole Earth, the whole Circuit, and whole Extent of it, burying all in Water, even the greatest Mountains*; which is, in plain Terms, such a one as was *explain'd and understood* by *Moses*, and the *Generality* of Writers since.

Having therefore thus over-hastily concluded that such a *Deluge* was *impossible*: and that all Nature could not afford Water enough to drown the *whole Globe*, if of the Circuit and Extent that now it is; he flies to a new *Expedient* to solve the Matter, and supposes an *Earth* of a *Make*

Make and *Frame* much like that imaginary one of the famous Monſieur *Des Cartes**; which he fancies to fall all to Pieces, at the *Deluge*, and to contract it ſelf into a leſſer Room, that a leſs Quantity of Water might ſurround and encompaſs it.

* *Princ. Philoſ.* l. 4.

The ſober and better Sort of the Standers-by, and thoſe who were Well-wiſhers to *Moſes*, began to be under ſome Concern and Uneaſineſs to ſee him thus ſet aſide only to make Way for a *new Hypotheſis:* and ſo *ſerious* and *weighty* a *Matter*, as is this *Relation* of the *Univerſal Deluge*, plac'd after all upon ſo unſteady a *Bottom*. But that *Concern* encreas'd when they further heard him ſo zealouſly decrying all former *Notions* of a *Deluge:* refuſing to grant one upon any *Terms* but his own : and ſo peremptorily declaring, *That all other Ways aſſign'd for the Explication of Noah's Flood are falſe or impoſſible.* This was to reduce the Thing to a very great *Strait:* and ſurely an expoſing and venturing of it a little too far. For, if all the other *Ways* be

falſe

falſe and *impoſſible*, ſhould *this*, the only one left, prove at laſt ſo likewiſe, the *Opinion* of a *Deluge* would be left very precarious and defenſeleſs: and we might either believe or diſbelieve it at Pleaſure. Nay the *negative Part* would, of the two, have much the *Advantage*; there being no reaſonable *Foundation* to believe that the *Deluge* did come to paſs *this Way*.

Some Men there are who have made a very untoward Uſe of *this*: and ſuch a one as I am willing to perſuade my ſelf he never intended they ſhould; yet it were to have been wiſh'd that he had been ſomewhat more wary. Theſe cry'd up his *Computation* of the *Water* as indiſputable and infallible: and then boldly gave out that ſuch a *Deluge* as that deſcribed by *Moſes* was altogether incredible, and that there never was nor could be any ſuch Thing. Nothing was talk'd of amongſt them under Mathematical Demonſtrations of the *Falſhood* of it; which they vented with all imaginable Triumph, and would needs have it that they had here ſprung

ſprung a freſh and unanſwerable *Argument* againſt the *Authenticks* of the *Moſaick Writings*; which indeed is what they drive at, and a *Point* they very fain would gain.

For my Part, my *Subject* does not neceſſarily oblige me to look after this *Water*: or to point forth the *Place* whereunto 'tis now retreated. For, when, from the *Sea-Shells*, and other *Remains* of the *Deluge*, I ſhall have given undeniable *Evidence* that it did actualy cover all *Parts* of the *Earth*, it muſt needs follow that there was *then* Water *enough* to do it, wherever it may be *now* hid, or whether it be *ſtill* in *being* or not. Yet the more effectualy to put a Stop to the *Inſults* and *Detractions* of theſe vain *Men*, I reſolved to enter a little farther into the Examination of this Matter. Which produced the *former Section* of this 3d *Part*, wherein I enquire what *Proportion* the *Water* of the *Globe* bears to the *Earthy Matter* of it. And upon a moderate *Eſtimate* and *Calculation* of the *Quantity* of *Water* now actualy contain'd in the *Abyſs*, I found that *this alone* was *more than* *enough*

enough, if brought out upon the *Surface* of the *Earth*, to cover the *whole Globe* to the Height aſſign'd by *Moſes*; which is *fifteen Cubits* above the *Tops* of the *higheſt Mountains**. The Particulars of which *Calculation*, ſhall be laid before the Reader at *Length* in the *Larger Work*. For any one will eaſily ſee that there is ſo great an *Apparatus* of Things, only *previous*, which muſt needs be *adjuſted* before I can come to the *Calculation* it ſelf, that to deſcend to Particulars *here*, further than I have already * done, would not only carry *this Diſcourſe* out beyond all reaſonable *Bounds*, and make the *Parts* of it *diſproportionate* to each other, but, which is not leſs to be thought of, would be an *Anticipation* of the *Other Work*.

* Gen. vii. 20.

* *Confer Sect.* 1. *Conſ.* 12.

This done, I again ſet aſide the *Obſervations* about the *Fluids* of the *Globe*, introduced upon this *Occaſion* in the *other Section*, as now of no farther Uſe: and *reaſſume* the *Thread* of the *other Obſervations* which I propoſe at the Beginning of *this Work*: and from them I ſhew,

1. That the *Deluge* was *Univerſal*, and laid the *whole Earth* under *Water*; covering all, even the *higheſt*, *Mountains*, quite round the *Globe*.

2. That, at the Time of the *Deluge*, the *Water* of the *Ocean* was *firſt* born forth upon the *Earth*: that it was immediately ſucceeded by *that* of the *Abyſs*; which likewiſe was brought out upon the *Surface* of the *Globe*.

3. That upon the *Diſruption* of the *Strata*: and the *Elevation* of ſome, and *Depreſſion* of others of them, which follow'd after that *Diſruption*, towards the latter End of the *Deluge**, this Maſs of *Water* fell back again into the *depreſt* and *lower Parts* of the *Earth*: into *Lakes* and other *Cavityes*: into the *Alveus* of the *Ocean*: and, through the *Fiſſures* whereby *this* communicates with the *Ocean**, into the *Abyſs*; which it filled till it came to an *Æquilibrium* with the *Ocean*.

* *Part* 2. *Conſect.* 6.

* *Sect.* 1. *ſupra. Conſect.* 2.

4. That there muſt have paſs'd a conſiderable Number of *Years* betwixt the *Creation* and the *Deluge*: and moſt probably about ſo many as *Moſes* hath aſſign'd.

That

That the *Deluge* commenc'd in the *Spring-season*; the *Water* coming forth upon the Earth in the *Month* which we call *May**. 5.

* Confer p. 81, 82 supra: & Part 6. sub finem.

That not only *Men*, *Quadrupeds*, *Birds*, *Serpents*, and *Insects*, the Inhabitants of the Earth and Air: but the far *greatest Part* of all Kinds of *Fish* likewise, the Inhabitants of the Sea, of Lakes, and of Rivers, *suffer'd* under the Fury of the *Deluge*, and were *kill'd* and *destroy'd* by it. 6.

That the *Deluge* did not happen from an accidental Concourse of *Natural Causes*, as the *Author* above-cited is of Opinion*. That very many *Things* were then certainly done, which never possibly could have been done without the Assistance of a *Supernatural Power*. That the said *Power* acted in this Matter with *Design*, and with the highest *Wisdom*. And that, as the *System* of *Nature* was *then*, and is *still* supported and establish'd, a *Deluge* neither could *then*, nor can *now*, happen *naturaly*. 7.

* Theory of the Earth. l. 1. c. 6, 8, &c.

I close up this Section with *two additional Discourses*,

N 2 The

The firſt concerning the *Migration of Nations*; with the ſeveral *Steps* whereby the *World* was *re-peopled* after the *Deluge* by the *Poſterity* of *Noah*, and particularly that mighty Tract of *America*. Wherein I ſhall make out, 1. *Who* they were that firſt peopled it. 2. *When* they departed thitherwards. 3. *What Courſe* they took: and by *what Means* both *Men* and *Beaſts*, as well as *Serpents* and the other noxious and more intractable Kinds of them, as the more innocent and uſeful, *got thither*. 4. Whether there remain any certain *Veſtigia* of a *Tradition*, in the *Writings* of the *Antients*, about theſe *Americans*: and what Country they intended under the Name of *Atlantis*. 5. Whether the *Phœnicians*, or any other Nation of the *old World*, maintained antiently any *Commerce* or *Correſpondence* with them. 6. How it happen'd that both the Inhabitants of *that*, and of *our World*, loſt all *Memory* of their *Commigration* hence. 7. Whence came the *Difference* in *Perſon*, or in the external *Shape* and *Lineaments* of the *Body*: in *Language*: in *Dyet*, and

Manner of *Living*: in *Clothing*: in *Arts* and *Sciences*: in *Customs*, Religious, Civil, and Military, betwixt these *Americans*, and their old Relations in *Asia*, *Europe*, and *Africa*. With Animadversions on the Writings of *Grotius*, *De Laet*, *Hornius*, and others, upon *this Subject*.

The Second concerning the unanimous *Tradition* of an *Universal Deluge* amongst all the most *antient Gentil Nations*; particularly the *Scythians*, the *Persians*, the *Babylonians*: the *Bithynians*, *Phrygians*, *Lydians*, *Cilicians*, and other People of *Asia Minor*: the *Hierapolitans*, *Phænicians*, and other Inhabitants of *Syria*: the *Ægyptians*, *Carthaginians*, and other *African* Nations: the most antient Inhabitants of the several Parts of *Greece*: and of the other Countryes of *Europe*: the old *Germans*: the *Gauls*: the *Romans*: the antient Inhabitants of *Spain*: and even the *Britains* themselves, the first Inhabiters of this Island; proving that the great Devastation and Havock the *Deluge* made, both of the *Earth* it self, of the Generality of *Mankind*, of *Brutes*,

 and

and all other *Animals*, had wrought a deep and very ſenſible *Impreſſion* upon the Minds of theſe *ancient Nations*. who lived nearer to the *Time* of it. That they had not only a *Memory* and *Tradition* of it in *general* and *at large*: but even of ſeveral the moſt remarkable particular *Accidents* of it likewiſe; which they handed downwards, to the *ſucceeding Ages*, for ſome time, with Notes of the greateſt Terror Amazement and Conſternation expreſſible. That it was commemorated chiefly by certain *Religious Rites* and *Ceremonies* uſed by them in the *Worſhip* of the EARTH. Which Superſtitious *Adoration* was firſt inſtituted upon *this Occaſion*, in thoſe ſimple and ignorant Ages*, and addreſs'd to the *Earth*, not only *expreſly* and by *Name*, but alſo under the feigned and borrow'd Names of *Atergatis*, *Derceto*, *Aſtarte*, *Dea Syria*, *Herthus*, *Iſis*, *Magna Mater*, *Cybele*, and *Rhea*, with ſeveral more; by all which, they intended the *Earth*. That at length the *Tradition*, for want of *Letters*, which were not then *invented*, or ſome

* *Vid. pag.* 59, & *ſeqq. ut* & *pag.* 106 & 107.

ſome other like Means to *preſerve* it, *wearing* out: and the *Reaſon* of the Inſtitution of this *Worſhip* being by Degrees *forgot*, the *After-Ages* perverted it to a ſomewhat different *Senſe* and *Intention*; ſuppoſing that this was only a reverential *Duty* and *Gratitude* paid to the *Earth* as the common *Parent* of *Mankind*, and becauſe both *Man*, and all other *Creatures* proceed out of it. By which Means the *true Notion* of the *Inſtitution* being *loſt*, the *Tradition* of the *Deluge*, which was couched under it, was alſo thereupon at length *ſuſpended* and *loſt*; few or none of all theſe many *Nations*, in the *latter Ages* of the *World*, having any Memory or Knowledge of it, beſides what they afterwards recovered from the *Jews* and *ancient Chriſtians*, who had it from the Writings of *Moſes*. In *Greece* indeed there were ſome other *Accidents*, which perplex'd and impeded the *Tradition* of it in that Country, whereof I have already * given ſuch Hints as this Diſcourſe will bear.

* *Part* I. p. 72, 73.

PART IV.

Of the Origin and Formation of Metalls *and* Minerals.

WHAT I can advance, with competent Certainty, about the *Fluids* of the *Globe*, the Sea, Springs, Rivers, and Rain, I propoſe in the immediately foregoing or Third Part of this *Eſſay*. As in the Second Part of it I diſpatch the *Solids*; Stone, Marble, Clay, and all the other *Terreſtrial Matter* of it, which is digeſted into *Strata*. That Part therefore comprehends the far greater Share of that Matter: and indeed all, excepting only *Metalls* and *Minerals*; which are found much more ſparingly, and in leſſer Parcels; being either *encloſed* in thoſe *Strata* (lying amongſt

amongſt the Sand, Earth, or other Matter whereof they conſiſt) or contain'd in their *perpendicular Fiſſures.* And *theſe* remaining ſtill to be conſider'd, I have allotted this *Fourth Part* to that Purpoſe.

To write of *Metalls* and *Minerals* intelligibly and with tolerable Perſpicuity, is a Taſk much more difficult than to write of either *Animals* or *Vegetables.* For thoſe carry along with them ſuch plain and evident *Notes* and *Characters* either of Diſagreement, or Affinity with one another, that the ſeveral *Kinds* of them, and the ſubordinate *Species* of each, are eaſily known and diſtinguiſh'd, even at firſt Sight; the Eye alone being fully capable of judging and determining their mutual *Relations*, as well as their *Differences.*

But in the *Mineral Kingdom* the Matter is quite otherwiſe. Here is nothing *regular*, whatever ſome may have pretended: nothing *conſtant* or *certain.* Inſomuch that a Man had need to have all his Senſes about him: to uſe repeated Tryals and Inſpections, and that with all

all imaginable Care and Waryneſs, truly and rightly to diſcern and diſtinguiſh Things, and all little enough. Here are ſuch a vaſt Variety of *Phænomena* : and thoſe, many of them, ſo deluſive, that tis very hard to eſcape Impoſition and Miſtake. *Colour*, or outward Appearance, is not at all to be truſted. A common *Marcaſite* or *Pyrites* ſhall have the Colour of *Gold* moſt exactly : and ſhine with all the Brightneſs of it ; and yet, upon Tryal, after all, yield nothing of Worth, but *Vitriol*, and a little *Sulphur* ; whilſt another Body, that hath only the Reſemblance of an ordinary *Peble*, ſhall yield a conſiderable Quantity of *Metallick* and valuable Matter. So likewiſe a Maſs, which, to the Eye, appears to be nothing but meer ſimple *Earth*, ſhall, to the Smell or Taſte, diſcover a plentifull Admixture of *Sulphur*, *Alum*, or ſome other *Mineral*.

Nor may we with much better Security rely upon *Figure*, or external *Form*. Nothing more uncertain and varying. 'Tis uſual to meet with the very ſame *Metall*, or *Mineral*,

Mineral, naturaly ſhot into quite *different Figures*: as 'tis to find quite *different Kinds* of them all of the *ſame Figure*. And a Body that has the Shape and Appearance of a *Diamond*, may prove, upon Examination, to be nothing but *Cryſtall*, or *Selenites*: nay perhaps only common *Salt*, or *Alum*, naturaly cryſtalliz'd and ſhot into that *Form*.

So likewiſe if we look into their *Situation*, and *Place* in the Earth; ſometimes we find them in the *perpendicular Intervalls*: ſometimes in the *Bodyes* of the *Strata*, being interſpers'd amongſt the *Matter* whereof they conſiſt: and ſometimes in both. Even, if I may ſo ſpeak, the *gemmeous Matter* it ſelf; with this only Difference, that thoſe *Gemms*, *e. g.* Topazes, Amethyſts, or Emeralds, which grow in the *Fiſſures*, are ordinarily *cryſtalliz'd*, or ſhot into *angulated Figures*: whereas, in the *Strata*, they are found in *rude Lumps*, and only like ſo many yellow, purple, and green *Pebles*. Not but that even theſe, that are thus lodg'd in the *Strata*, are alſo ſometimes

times found *cryſtalliz'd*†, and in Form of *Cubes*, *Rhombs*, and the like*. Or if we have reſpect to the *Terreſtrial Matter wherein they lye* in thoſe *Strata*, here we ſhall meet with the ſame *Metall* or *Mineral* embody'd in *Stone*, or lodg'd in *Cole*, that elſewhere we found in *Marle*, in *Clay*, or in *Chalk**.

* *Vid. Conſect.* 2. *infra.*

* *Vid. Conſect.* 3. *infra.*

As much Inconſtancy and *Confuſion* is there in their *Mixtures with each other*, or their Combinations amonſt themſelves. For 'tis rare to find any of them *pure*, *ſimple*, and *unmixt*;

† The *Cryſtallized Bodyes* found in the *perpendicular Intervalls* are eaſily known from thoſe which are lodged in the *Strata*, even by one who did not take them thence, or obſerve them there. The *former* have always their *Root*, as the Jewellers call it; which is only the *Abruptneſs* at that *End* of the Body whereby it adhered to the Stone, or Sides of the Intervalls; which *Abruptneſs* is cauſed by its being broke off from the ſaid Stone. Thoſe which are found in the *Strata* of Earth, Sand, or the like, (having lain *looſe* therein) are *intire*, and want that Mark of Adheſion. But thoſe which are *incloſed* in Stone, Marble, or ſuch other *ſolid Matter*, being difficultly ſeparable from it, becauſe of its Adheſion to all Sides of them, have commonly ſome of that Matter ſtill adhering to them, or at leaſt Marks of its having been broke from them, on *all their Sides*; wherein theſe differ from thoſe found in the *perpendicular Intervalls*, they adhering, as we have noted, by only *one End*. *Vid. Conſ.* 6. *&c. infra.*

unmixt; but Copper and Iron together in the *ſame Maſs*: Copper and Gold: Silver and Lead: Tin and Lead: yea ſometimes all the ſix *promiſcuouſly* in *one Lump*. 'Tis the ſame alſo in *Minerals*; Nitre with Vitriol: Common Salt with Alum: Antimony with Sulphur: and ſometimes all theſe together. Nor do *Metalls* only ſort and herd with *Metalls* in the Earth: and *Minerals* with *Minerals*; but *both* indifferently and in common *together*. Lead, with Spar, with Calamin, or with Antimony: Iron with Vitriol, with Alum, with Sulphur: Copper with Sulphur, with Vitriol, *&c.* yea Iron, Copper, Lead, Nitre, Sulphur, Vitriol, and perhaps ſome more, in one and the *ſame Maſs*. In a Word, the only ſtanding *Teſt*, and diſcriminative *Characteriſtick* of any *Metall* or *Mineral* muſt be ſought for in the *conſtituent Matter* of it: and it muſt be firſt brought down to *that* before any certain *Judgment* can be given. And when that is once done, and the ſeveral Kinds *ſeparated* and *extracted* each from the other, an homogeneous Maſs of

one

one Kind is eaſily diſtinguiſhable from any other: Gold from Iron: Sulphur from Nitre: and ſo of the reſt. But, without *this*, ſo various are their *Intermixtures*, and ſo *different* the *Face* and *Appearance* of each, becauſe of that *Variety*, that ſcarcely any thing can be certainly determin'd of the particular *Contents* of any ſingle *Maſs* of *Ore* by meer *Inſpection*. I know that by *Experience* and *Converſation* with theſe *Bodyes*, in any *Place* or *Mine*, a Man may be enabled to give a near Conjecture at the *Metallick* or *Mineral Ingredients* of any *Maſs* commonly found there; but this meerly becauſe he hath before made Tryal of *other like Maſſes*, and thereby learned what it is they contain. But, if he remove to *another Place*, tho' perhaps very little diſtant, 'tis ten to one but he meets with ſo different a *Face* of *Things*, that he'll be there as far to ſeek in his *Conjectures* as one who never before ſaw a *native Ore*; untill he hath here made his *Tryals* as before, and ſo further inform'd himſelf of the Matter.

Metalls

Metalls being ſo very uſeful and ſerviceable to *Mankind*, great Care and Pains hath been taken, in all Ages, in *Searching* after them, and in *Separating* and *Refining* of them. For which Reaſon 'tis that *theſe* have been accurately enough *diſtinguiſh'd* and reduced to *ſix Kinds*; which are all well enough known. But the like Pains hath not been taken in *Minerals*; and therefore the Knowledge of *them* is ſomewhat more confuſed and obſcure. *Theſe* have not yet been well reduced, or the *Number* of the *ſimple original* ones rightly fixt; *ſome*, which are only *Compounds*, the Matter of two or more *Kinds* being *mix'd* together, and, by the different *Proportion* and *Modulation* of that Matter, variouſly *diſguis'd* and *diverſify'd*, having been reputed all *different Kinds* of *Minerals*, and thereby the *Number* of them unneceſſarily *multiply'd*. Of this we have an *Inſtance* in the *Gemm-kind*; where, of all the *many Sorts* reckon'd up by *Lapidaries*, there are not above *three* or *four* that are *Originals*; their *Diverſities*, as to *Luſtre*, *Colour*, and *Hardneſs*, ariſing

rising from the different *Admixture* of other *adventitious Metallick* and *Mineral Matter.* But the farther and clearer Adjustment of *this Affair* I am constrained to adjourn to the *larger Treatise.*

In the mean time 'tis sufficient for my *present Design* to remark, in general, that those *Minerals* and *Ores* of *Metalls* which are reposited in the *Bodyes* of the *Strata*, are either found in *Grains*, or small Particles, dispersedly *intermix'd* with the *Corpuscles* of *Earth*, *Sand*, or other *Matter* of those *Strata*: or else they are *amass'd* into *Balls*, *Lumps*, or *Nodules.* Which *Nodules* are either of an *irregular* and *uncertain Figure*, such as are the common *Pyritæ*: *Flints*, *Agates*, *Oxyx's*: *Pebles*, *Cornelions*, *Jaspers*, and the like: or of a *Figure* somewhat more *regular* and *observable*, such as the *Belemnites*: the several Sorts of *Mineral Corall*, of the *Stelechites*, and of the *Lapis Mycetoides* *: the *Astroites*, or *Starry-Stone*, as well that Sort with the *Prominent*, as that with the *Plane*, and that with the *Concave Stars*: the *Selenites*: the *Echinated*

* *Vulgarly call'd* Fungites.

Echinated Cryſtalline Balls, with many more analogous Bodyes.

Thoſe which are contain'd in the *Perpendicular Intervalls* of the *Strata* are, either ſuch as are there accumulated into a *rude Heap*, without any particular *Form* or *Order*, being only included betwixt the two *oppoſite Walls* or *Sides* of the ſaid *Intervalls*, which they wholey or partly *fill*, as there is a greater or leſs *Quantity* of them; in which Manner *Spar* is uſualy found herein, and other *Minerals*, as alſo the common *Ores* of *Lead*, *Tin*, *Iron*, and other Metalls: or elſe ſuch as are of ſome *obſervable Figure*. Of this *Sort* are the *Sparry Striæ*, or *Iceycles*, called *Stalactitæ**: the *Native Saline Iceycles*, or *Sal Stalacticum*: the *Vitriolum Stalacticum nativum*: the *Vitriolum capillare*: the *Alumen Stalacticum*, and *capillare*: *Minera ferri Stalactica*, which, when ſeveral of the Cylindrick *Striæ* are contiguous, and grow together into one *Sheaf*, is called *Bruſh-Iron-Ore*: and laſtly the *Argentum arboreſcens*, *& capillare*. Hither alſo ought to be refer'd the *Cryſtallized Ores, and*

* *Or rather* Stagonitæ.

Minerals, e.g. the *Iron-Rhombs*, the *Tin Grains*: the *Mundick Grains*: the tessellated *Pyritæ*, or *Ludus Paracelsi*: *Crystallized Native Salt*, *Alum*, *Vitriol*, and *Sulphur*. As likewise the *Gemms* or *Stones* that are found in these *perpendicular Intervalls*, shot into *Cubes*, into *Pyramidal Forms*, or into *angulated Columns*, consisting sometimes of *three*, but most commonly of *six Sides*, and mucronated or terminating in a *Point*; being either *opake*, or *pellucid*: or partly pellucid, and partly opake, and *colour'd*, black, white, grey, red, purple, blew, yellow, or green; *e. gr.* *Crystall*, the *Pseudo-Adamantes*, the *Cornish-Stones*, the *Bristow Stones*, *Crystallized*, *Sparrs*, the *Amethyst*, the *Saphire*, the *Topaz*, the *Emerauld*, and several others.

My Business *here* is to enquire into the *Origin* and *Production* of these *metallick* and *mineral Bodyes*: to enquire how they came into *this* *Condition*, and attained *these Figures*. And as my *Observations* have been the *Light* whereby I have hitherto steered my *Course*, so I here betake my self unto *them* again; and 'tis from *them* that I prove,

That

That as the more *groſs* and *maſſive* Parts of the *Terreſtrial Globe*, the *Strata* of Stone, Marble, Earth, and the reſt, owe their *preſent Frame* and *Order* to the *Deluge**: ſo likewiſe do *Metalls* and *Minerals*; the far *greater Part* of them, I mean *all thoſe* which we now find lodged in theſe *Strata* amongſt the Sand, Earth, *&c.* being actualy repoſed therein during the Time that the *Water* covered the *Earth*: and the *Earth* it ſelf then put into ſuch a *Condition* that the *reſt*, I mean thoſe we now find in the *perpendicular Intervalls*, ſhould be *collected* thither *by Degrees*, and be *formed*, almoſt of Courſe, meerly by the *ordinary Motion* of the *Water*, and its *Paſſage* to and fro in the *Earth**. 1.

* *Vid. Part* II.

* *Vid. Conject.* 4. & 5. *infra.*

That whilſt the *Corpuſcles* of *Metalls* and *Minerals*, together with thoſe of Stone, Marble, Cole, Chalk, and the like *coarſer Matter*: as alſo the *Shells*, *Teeth*, and other Parts of *Animals*, and *Vegetables*, were *ſuſtained* in the *Water*, at the *Deluge**; after ſome Time, that the *Commotion* was over, and the *Water* come to a calm and ſedate State, 2.

* *Part* II. *Conject.* 2.

ſuch of thoſe *Corpuſcles*, as happen'd to *occur* or *meet* together, *affix'd* to *each other*: and, many of them convening, uniting, and combining into *one Maſs*, formed the *metallick* and *mineral Balls* or *Nodules* which we now find.

That all *metallick* and *mineral Nodules* whatever: as well thoſe which are in *rude Lumps*, ſuch as the common *Pyritæ*, *Flints*, *Agates*, *Onyxes*, *Pebles*, *Jaſpers*, *Cornelions*, and the like: as thoſe which are of a more *regular* and *obſervable Shape*, ſuch as the *Selenites*, *Belemnites*, *Aſtroites*, *Stelechites*, *mineral Coral*: and, in one Word, all *others* whatſoever, were *formed* at *this Time*, and by *this Means*.

That in *ſuch Parts* of the *Water* where the *Corpuſcles* ſo ſuſtain'd chanced to be all of the *ſame Kind**: or, at leaſt, where there were fewer Kinds or Varieties of them, the *Nodules*, which were thus form'd out of them, were more *ſimple*, *pure*, and *homogeneous*; as are the *Selenites*, and ſome Kinds of *Pebles* and *Flints*, to name no more. But where (as indeed it generaly fell out) there

* *Vid.* *Part* V. *Conſect.* 1.

there happened to be a *greater Variety* of *Corpuſcles*, as ſuppoſe of Nitre, of Vitriol, of Iron, of Copper, or whatſoever elſe, ſuſtained *promiſcuouſly* together, there the *Nodules*, formed out of them, were *mixt*, and conſiſted of a *greater Variety* of *Matter* confuſedly aſſociated into the *ſame Lump*. Of this the *Pyritæ* may ſerve for an *Example*; whereof ſome yield Iron, Sulphur, and Vitriol: others Copper and Alum, yea ſome of them contain all theſe, and ſeveral more, in the *ſame Nodule*.

That the *Bones*, *Teeth*, *Shells*, and other *like Bodyes*, being *ſuſtain'd* in the *Water* together with theſe *metallick* and *mineral Corpuſcles* *, and the ſaid *Corpuſcles* meeting with, and hitting upon *thoſe Bodyes*, they *affix'd* unto them, and became *conjoyned* with them; ſome of them (though this very rarely) paſſing into their *Pores* and *Interſtices*: others *adhering* in *Lumps*, or *Maſſes*, to their *Outſides*, and indeed oftentimes combining in ſuch Numbers upon the *exteriour Surface* of the *Shell*, *Tooth*, &c. as wholey to *cover* and

* *Part* II. *Conſect.* 2.

involve it in the *Mass* they together *constituted* : and others of them *entering* into the *Cavities* of the *Echini Cochleæ*, *Conchæ*, and other *Shells*, till they had quite *filled* them up; those *Shells*, by that Means, serving as *Proplasmes*, or *Moulds*, to the *Matter* which so *filled* them, limiting and determining both the *Dimensions* and *Figure* of it. That accordingly we at this Day find some few of these *fossil Shells*, and other *Animal-Substances*, with *Iron-Ore*, *Spar*, *Vitriol*, *Sulphur*, and the like, intruded into their *Pores*. But far greater Numbers of them with *Lumps* of *Flint*, *Ores* of *Metalls*, and *Minerals*, growing firmly to the *Outsides* of them, and oftentimes in such *Quantity* that the *Shell* or *Tooth* is wholey *covered* by those *Minerals*, being *immers'd* or *included* in the *Mass* they *constitute*. Insomuch that 'tis very usual, upon breaking *Flints*, *Pyritæ*, *&c.* to find *Pectines*, *Conchæ*, and the like, enclos'd, even in the very *Middle* of them. As common is it to find *Echini*, *Cochleæ*, *Conchæ*, and other *Shells*, having their *Cavities fill'd* up with

with *Ores* of *Metalls*, *Flint*, *Spar*, *Native Vitriol*, *Arſenic*, and other *Minerals*. Not but that theſe *Minerals* many times *ſurvive* the *Shells* which gave them their *Forms*, and are found even after *thoſe* are *rotted* and *diſappeared*. For tho', when lodg'd in *Chalk*, or the like *cloſe Matter*, which *preſerves* and *ſecures* them againſt *external Injuries*, thoſe *Shells* are conſtantly found upon, and actualy *inveſting* the *Flint*, *Spar*, or other *Mineral*, and are commonly as *fair* and *entire* as any of their fellow *Shells* at Sea; yet, when they happened to be lodg'd amongſt *Sand*, *Gravel*, or the like *laxer Matter*, the *Shells* are uſualy *periſh'd* and *gone* *, and ſo the *Flint*, *Spar*, *&c.* left *uncover'd*. In which Caſe, the ſaid *Flint*, *Spar*, or other *Mineral*, is of a *conſtant*, *regular*, and *ſpecifick Shape*, as is the *Shell* whence it borrows both that *Shape*, and indeed its *Name*; theſe being the *Bodyes* which are call'd, by Naturaliſts, *Echinitæ*, *Cochlitæ*, and *Conchitæ* *, as *reſembling* the *Shells* of thoſe Names. Which truly many of them do very nearly;

* *Vid.* *Part* III. *Sect.* 1. *Conſ.* 11.

* *Confer* *Part* V. *Conſect.* 5.

 they

they having taken the *Impresses* of the *Insides* of these *Shells* with that exquisite *Nicety* as to express even the smallest and finest *Lineaments* of them. Insomuch that no *Metall*, when melted and cast in a *Mould*, can ever possibly represent the Concavity of that *Mould* with greater *Exactness* than these *Flints*, and other *Minerals*, do the *Concavities* of the *Shells* wherein they were thus *moulded*.

3. That at length all this *metallick* and *mineral Matter*, both that which continu'd *asunder*, and in *single Corpuscles*, and that which was *amass'd* and *concreted* into *Nodules*, *subsided* down to the Bottom; at the same time that did the *Shells*, *Teeth*, and other like *Bodyes*: as also the Sand, Cole, Marle, and other *Matter* whereof the *Strata* of Sand-Stone, Cole, Marle, and the rest are for the most part *composed**; and so were *included in*, and *lodged* amongst, *that Matter*.

* *Confer Part* II. *Cons.* 3.

That in regard that both the *ordinary Terrestrial Matter*, and the *mineral* and *metallick Matter*, which was assumed up into the *Fluid* was *different*

different in *different Parts* of it; being in *ſome Places* all chiefly of *one Kind*, ſuppoſe *Sand:* in *others* of a *different Kind, e. gr. Chalk:* and in *others* of *ſeveral Kinds* together, as *Sand, Chalk*, and many more: and there being no other *Place* or *Apartment* in the *Globe* aſſigned to any of this *Matter* than *that* whereinto its own natural *Gravity* bore it, which was only *directly downwards*, whereby it obtained *that Place* in the *Globe* which was juſt underneath *that Part* of the *Fluid* where it was *ſuſtained* when the *Subſidence* began; it thence happened that the *Strata*, which were afterwards *conſtituted* by this *Matter* thus *ſubſiding*, are alſo *different* in *different Places:* in ſome all, or moſt of them, of *Sand-ſtone:* in others of *Chalk:* and in others of both *Sand-ſtone* and *Chalk*, and perhaps *many more*, lying each upon other. And the Caſe of *Metalls* and *Minerals* being the ſame, 'tis for that Reaſon that in ſome Places we now get *Iron*, or *Vitriol*, but no *Copper*, or *Alum:* in others we find *theſe*, but not *thoſe:* and, in others, both *theſe*, and *thoſe*, and perhaps *many more*.

That

That the *Place*, both of the several Sorts of *Terrestrial Matter*, and of the *Metalls*, and *Minerals*, whilst *sustain'd* in the *Fluid*, being thus *contingent*, and *uncertain*, their *Intermixtures* with each other, and with the *Terrestrial Matter*, in the *Sediment*, or *Strata*, which *subsiding* they together *composed*, must consequently be *uncertain* likewise; that *Metall*, or *Mineral*, of whatever *Kind* it chanced to be, which was *sustain'd* in *any Part* of the *Fluid*, *setling* only directly downwards, was *lodged* amongst the *Terrestrial Matter* which chanced to be *sustained* together with it in the *same Part*, of what Kind soever that *Matter* was. And accordingly we now find them *uncertainly* mixt; the same *Metall* or *Mineral* lodged, in *some Places*, in *Stone*: in *others*, in *Cole*: and in *others*, in *Clay*, *Marle*, or any other *Matter* indifferently*. And as we find the *same Metall* or *Mineral* lodged amongst *different Sorts* of the common *Terrestrial Matter*, so do we, for still the same Reason, also find *different Kinds* of *Metalls* and *Minerals*, *Copper*, *Iron*, and

* *Vid.* *Pag.* 192. *supra.*

and *Sulphur*, *Tin*, *Lead*, and *Vitriol*, with *ſeveral more*, lodged all confuſedly together in the very *ſame Sort* of *Terreſtrial Matter* *.

* *Vid. Pag.* 193. *ſupra.*

That the *Quantity* of the *metallick* or *mineral Matter* taken up into the *Fluid* was *various* and *uncertain*; there being in *ſome Parts* little, or perhaps none, of it: in *others* a very great Abundance. And ſo we at this Day find it; in *ſome Places*, little, or none: in *others*, in ſuch Plenty as to exceed even the *ordinary Terreſtrial Matter*, and of it ſelf to compoſe whole *Strata*, without any conſiderable *Admixture* of Sand, Clay, or other *common Matter*. Thus we ſometimes ſee whole *Strata* compiled of *metallick* and *mineral Nodules*: others of *Pebles*, and of *Flints*, without the Interpoſition of other Matter; that *finer Matter*, commonly found amongſt theſe, and vulgarly called *Sand*, being realy no other than very *ſmall Pebles*, as may appear to any one who ſhall carefully examine and obſerve it, eſpecialy with a good *Microſcope*. Thus likewiſe we find *Strata* conſiſting almoſt entirely of *Common Salt*:

Salt: others of *Ochre:* and others of ſeveral *Metalls* and *Minerals*, *Tin*, *Lead*, *Vitriol*, *Nitre*, and *Sulphur*, promiſcuouſly, without any conſiderable *Intermixture* of coarſer *Terreſtrial Matter*.

4. That the *metallick* and *mineral Matter*, which is *now* found in the *perpendicular Intervalls* of the *Strata*, was all of it *originaly*, and at the Time of the *Deluge*, lodged in the *Bodyes* of thoſe *Strata*; being interſperſed or ſcatter'd in *ſingle Corpuſcles*, amongſt the *Sand*, or other *Matter*, whereof the ſaid *Strata* mainly *conſiſt*. That it was *educed* thence, and *tranſmitted* into theſe *Intervalls*, *ſince* that Time; the *Intervalls* themſelves not *exiſting* till after the *Strata* were formed, and the *metallick* and *mineral Matter* was actualy lodged in them; they being only *Breaches* of the *Strata*, and not made till the very Concluſion of the *Cataſtrophe*, the *Water* thereupon immediately withdrawing again from off the *Earth**.

* *Confer Part* II. *Conſ.* 3. & 6. *Part* III. *Sect.* 2. *Conſect.* 3. & *Pt.* IV. *Conſect.* 3.

5. That the *Water*, which *aſcends* up out of the *Abyſs*, on all Sides of the *Globe*, towards the *Surface* of

of the *Earth* *, inceſſantly *pervading* the *Pores* of the *Strata*, I mean the *Interſtices* of the *Sand* or other *Matter* whereof they conſiſt, *detaches* and *bears* along with it all ſuch *metallick*, *mineral*, and other *Corpuſcles* which lye *looſe* in its Way, and are withal ſo *ſmall* as to be able to *paſs* thoſe *Interſtices*; forcing them along with it into the *perpendicular Intervalls*; to which it naturaly directs its *Courſe*, as finding there a ready *Exit* and *Diſcharge* *, being partly *exhaled* thence up into the *Atmoſphere*, and partly *flowing forth* upon the *Surface* of the *Earth*, and forming *Springs* and *Rivers*.

* *Part* III. *Sect.* 1. *Conſect.* 8.

* *Ibid.*

That the *Water* which falls upon the *Surface* of the *Earth* in *Rain*, bears alſo ſome, tho' a leſſer, Share in this *Action*; this, ſoaking into the *Strata* which lye near the *Surface*, ſtraining through the *Pores* of them, and *advancing* on towards their *perpendicular Intervalls* *, bears thither along with it all ſuch moveable *Matter* as occurrs in thoſe *Pores*, in much the ſame Manner as does the *Water* which ariſes out of the *Abyſs*; with this only *Difference*, that

* *Part* III. *Sect.* 1. *Conſect.* 4.

that *this* passes and pervades none but the superficial and uppermost *Strata*, whereas *the other* permeates also those which lye lower and deeper.

That the *metallick* and *mineral Corpuscles*, being thus *convey'd* into these *Intervalls*: and the *Water* there having more *Room* and freer *Passage* than before, whilst it only penetrated the *Pores* of the *Stone*, it *deserts* the said *Corpuscles*, leaving them in these *Intervalls*; unless it flow *forth* with a very *rapid* and *precipitate Motion*; for then it *hurries* them out along with it, 'till its *Motion* becomes more *languid* and *remiss*, when it *quits* and *abandons* them *.

* *Confer Cons.* 12. & 13. *infra.*

6. That by the *Water*, thus passing through the *Stone* to its *perpendicular Intervalls*, was brought thither all the *metallick* and *mineral Matter* which is now lodged therein: as well that which lyes only in an *indigested* and *confused Pile* *, in which Manner the far *greatest Part* of it is found, and particularly the *common Ores* of *Metalls Iron*, *Tin*, *Lead*, and the rest, as also *Spar*, and other *Minerals*:

* *Vid. Pag.* 197. *supra.*

Minerals: as that which is dispos'd and form'd into some *observable Figure*, such as the *metallick* and *mineral Stalactitæ*, the *angulated* or *Crystallized Metalls* and *Minerals* *, and to be short, all others whatever.

* *Vid. Pag.* 197. *supra.*

7. That there is not, whatever some Men may have fancy'd, any thing very *strange* and *extraordinary* in the *Production* of the said *formed Metalls* and *Minerals*, which are found in these *Intervalls:* nor other *plastick Vertue* concern'd in shaping them into those *Figures* than meerly the *Configurations* of the *Particles* whereof they consist, and the simple *Motion* of the *Water* to bring those *Particles* together. That particularly the common *Stalactites*, *Lapis Stillatitius*, or *Drop stone*, which consists principaly of *Spar*, and is frequently found, in Form of an *Icey-cle*, hanging down from the Tops and Sides of *Grotto's*, and of the lesser *perpendicular Intervalls*, was formed by the *Water* which continualy is passing through the *Strata* into these their *Intervalls*. For *this* takes the *Sparry Particles* as they lay dispersedly mingled with the *Sand*,

or

or other *Matter* whereof thoſe *Strata* conſiſt, and *bears* them on with it to the ſaid *Intervalls*; where iſſuing leiſurely out of the *Strata*, and having now free Paſſage, it *deſerts* theſe *Particles*, falling down from the *Tops* and *Sides* of the *Grotto's*, to which the *Particles* affixing by little and little, *incruſt* them over with a *Sparry Cover*, and alſo *form* theſe *Stalactitæ*, from which the *Water* is continualy falling and diſtilling Drop by Drop; which gave Occaſion to that *Miſtake* of thoſe who ſuppoſe theſe *Bodyes* to be only *Water* petrify'd, as they ſpeak, or converted into theſe *Sparry* or *Stoney Iceycles*, in the ſame Manner as it is by *Froſt* congeal'd into the *Icey* ones which hang down from the Eaves of Houſes, from Pipes, or other Conveyances of Water. That the *Iron*, and *other metallick Stalactitæ*: the *Aluminous*, and the *Vitriolic Stalactitæ*: the *Saline* ones, or thoſe which conſiſt of *common Salt*, and *all others*, are found *ſuſpended* in the *ſame Manner*, and their *Matter* was conducted out of the *Strata* to their Fiſſures by the *ſame Means*.

That

That the *Iron-Rhombs*, *Tin-Grains*, and other *Ores* of *Metalls*, which are found in these *Intervalls* naturaly formed into *Cubick*, *Pyramidal*, or other *Figures*: as likewise the *Minerals* which are there *shot* into the like *Figures*, such as the *tessellated Marcasits*, *crystallized native Salt*, *Alum*, *Vitriol*, and *Sulphur*: the *Gemms* also which are thus *figur'd*, *e. gr. Crystall*, the *Pseud-Adamantes*, the *Amethyst*, *Emerauld*, and the rest; I say *these*, and all other natural *metallick* and *mineral Crystallizations*, were *effected* by the *Water*, which first brought the Particles, whereof each consists, out from amongst the Matter of the *Strata*, into these their *Intervalls*, in much the same Manner that the *common*, or *artificial Crystallizations* of Alum, Vitriol, and the like, are now effected in the Water wherein they were before dissolved: and as are the *Chymical Crystallizations* of other *Minerals* and *Metalls* in their several *Menstrua*; whereof more in its Place. 8.

That the *Corpuscles* of *Metalls* and of *Minerals* being *smaller* than 9.

 those

thoſe of *Sand* and of the other *common Terreſtrial Matter*, and conſequently the *Pores* of the *Strata*, which conſiſt mainly, or at leaſt contain in them a conſiderable Quantity of *theſe**, being *leſſer* and *narrower* than *thoſe* of the *Strata* of *Sand-ſtone*, and the like common and *craſſer Matter*, the *Water* which aſcends from beneath towards the Surface of the Earth is admitted into them, if at all, only in *leſſer Quantity*, paſſes them *ſlowly* and *difficultly*, and therefore hath not *Scope* and *Power* ſufficient to *diſlodge* the *Corpuſcles*, and *bear* them off with it into the *perpendicular Intervalls*, as it does in thoſe *Strata* which conſiſt chiefly of *Stone*, and the like *groſſer Matter*, where the *metallick* and *mineral Corpuſcles* lye thinner, and ſo the *Pores* are more *wide* and *open*. That, for this Reaſon, in the *Intervalls* of thoſe *Strata* which abound plentifully with *Iron*, *Tin*, *Spar*, *common Salt*, *Alum*, or the like, we ordinarily find a *leſſer Quantity* of theſe *Metalls* and *Minerals* reſident, than we do in the *Intervalls* of ſome other *Strata* which now ſhew

* *Confer Conſect.* 3. *pag.* 207, 208, *ſupra.*

ſhew little, or perhaps nothing in the *Bodyes* of them beſides *Sand* and ſuch like *coarſer Matter.* For there is ſo admirable a *Contrivance* in this *Affair*, that the *Water* does not *diſturb* and *remove* that *metallick* or *mineral Matter* which lyes in the *Strata* in great *Plenty*, and ſo is there *ready collected* to the Hands of *Man:* but only *that* which needs ſuch an *Agent* to *collect* it: *that* which is ſo *ſparingly* and *diſperſedly* intermix'd with the *common terreſtrial Matter*, as not to be *diſcoverable* by *humane Induſtry*; or, if *diſcoverable*, ſo *diffuſed* and *ſcatter'd* amongſt the craſſer and more unprofitable *Matter*, that 'twould never be poſſible to *ſeparate* and *extract* it. Indeed if 'twas, it would not defray the *Charge* and *Labour* of the *Extraction:* and therefore it muſt needs have been all irretrievably *loſt*, and *uſeleſs* to *Mankind*, was it not here, by this Action of the *Water collected* and brought into one *Maſs*.

That therefore the *Metalls* and 10
Minerals which are lodged in the *perpendicular Intervalls* of the *Strata* do *ſtill grow*, (to ſpeak in the *Mine-*

raliſts Phraſe) or receive additional *Increaſe* from the *Corpuſcles* which are *yet daily* borne along with the *Water* into *them.* Nay they have *grown* in like Manner ever ſince the Time of the *Deluge*, in all ſuch *Places* where thoſe *Intervalls* are not already ſo *filled* that they cannot receive any more: or where the *Stock* of *metallick* and *mineral Corpuſcles*, originaly lodged in the *Strata*, is not quite *exhauſted*, and all borne thither already. That yet this *Increaſe* is not *now* any where *very great*; the *Corpuſcles*, which were capable of being *ſtirr'd* and *remov'd*, being, by the continual *Paſſage* of the *Water* for ſo many *Ages*, in moſt Places *exhauſted*, educed forth of the *Strata*, and *tranſmitted* into theſe their *Fiſſures.*

11. That the *metallick* and *mineral Matter* which lyes in the *Bodyes* of the *Strata* does *not* now *grow:* nor hath it ever receiv'd any *Addition* ſince 'twas firſt repoſed in thoſe *Strata* at the Time of the *Univerſal Deluge*; but, on the contrary, hath been *diminiſh'd* and *leſſen'd* by ſo much as hath been *convey'd* into their

their *perpendicular Intervalls*, and as hath been brought forth upon the *Surface* of the *Earth* by *Springs*, *Rivers* *, and *Exhalations* † from the *Abyſs*, ever ſince *that Time*. That notwithſtanding there have, and do ſtill happen, *Tranſitions* and *Removes* of it, in the *ſolid Strata*, from one Part of the ſame *Stratum* to another Part of it, occaſion'd by the *Motion* of the *Vapour* towards the *perpendicular Intervalls* of *theſe* *: and in the *laxer Strata*, ſuch as thoſe of *Sand*, *Clay*, and the like, from the *lower ones* to thoſe which lye *above* them, and even to the very *Surface* of the *Earth*, occaſioned by the *Motion* of the *Vapour* directly towards the *Surface*, it pervading theſe looſer *Strata* diametricaly *. But of this I have not Room to enlarge more particularly in this Place.

* *Vid. Conſ.* 12. *infra.*

† *Vid. Conſ.* 14. *infra.*

* *Part* III. *Conſect.* 8.

* *Ibid.*

12. That the *Bitumen* which is found in Lumps, or coagulated Maſſes, in ſome *Springs*: and which is, in others, found floating in Form of an *Oyl* upon the Surface of the Water, when 'tis called by Naturaliſts *Naphtha*, and *Petroleum*: the *Salt* where-

with the *Salinæ*, or *Salt-Springs*, abound: the *Vitriol*, *Alum*, *Nitre*, *Sulphur*, *Spar*, and other *Minerals*, wherewith the *Acidulæ*, or *Medicinal-Springs*, are ſaturated; I ſay, all theſe *Minerals* were *originaly* lodged in the *Strata* of Stone, Cole, Earth, or the like: that they were *educed* thence, and *convey'd* into theſe *Springs*, by the *Water* pervading thoſe *Strata* in its *Paſſage* from the *Abyſs* towards the ſaid *Springs**.

* Confer Part III. Sect. 1. Conſect. 8. & Pt. IV. Conſect. 5.

13. That when the *Water* of *Rivers* iſſues out of the Apertures of them, with more than ordinary *Agitation* and *Rapidity*, it uſualy *bears forth* along it ſuch *Particles* of *Spar*, *Argilla*, or other *looſe* and *moveable Matter*, as it met with in its *Paſſage* through the *Stone*, *Marble*, or other *ſolid Strata*. That it *ſuſtains* theſe *Particles*, and *carryes* them on together with it till ſuch Time as its *Motion* begins to *remitt* and be *leſs rapid* than it was at and near its *Sourſe*; when by Degrees it *lowers* them, and lets them *fall*, *depoſing* and *affixing* them upon any Thing which occurrs in the Way, as Stones, Shells, Sticks, or other like *Bodyes*; eſpecialy

eſpecialy thoſe which lye in the *Sinus's* or *Creeks* of thoſe *Rivers*, where the *Motion* of the *Water* is more *ſluggiſh* and *languid* than in the *Stream*, or Middle of the *Chanel.* That ſome *Rivers* do thus bring forth *Spar*, and other *mineral Matter*, in great *Quantity*, ſo as to *cover* and *incruſt* the Stones, Sticks, and other Bodyes lying therein, to a very conſiderable *Thickneſs.* That ſometimes the *Water* of *Standing-Springs* does the ſame; *precipitating* the *mineral Matter* which it brought forth of the *Strata*, upon the Stones at the Bottoms and Sides of the ſaid Springs: and *affixing* it upon Sticks, Straws, and other Bodyes, and upon the Moſs, or other Plants which happen to grow therein; *incruſting* them over, in like Manner as does the abovemention'd *Water* of *Rivers.*

That when the *Heat* at, and up- 14.
on, the *Surface* of the *Earth* is *great*, it not only *mounts up* the *Water* ſent from beneath, and, along with it, the lighter *Terreſtrial Vegetative Matter* *, but likewiſe the very *mineral Matter* it ſelf, *Sulphur*, *Nitre*, *Vitriol*,

* *Part* I. *Pag.* 50. & *Pt.* III. *Sect.* 1. *Conſect.* 8.

Vitriol, and the like; the *Atoms*, or *ſingle Corpuſcles* whereof, being *detach'd* from their reſpective *Beds* in the *Earth*, it *bears* quite to the *Surface* of it, and the *light* and more *active* Sorts of them up into the *Atmoſphere*, together with the *Vapour*, which, when condenſed, falls down again in *Rain*. That this Matter is thus carry'd up in greater or leſſer *Plenty*, and to a greater or leſſer *Height*, anſwerably to the greater or leſſer Quantity or Intenſeneſs of the *Heat*.

That wherever there happen to be any *extraordinary Diſcharges* of the *Subterranean Heat*; either *Vulcano's*, or leſſer *Spiracles*, ſuch as thoſe about *Naples*, *Pozzuolo*, and in other Parts of the World: *Thermæ*, or *Hot-ſprings*: or *firey Eructations*, ſuch as burſt forth of the *Earth* during *Earthquakes*; I ſay wherever there are ſuch or the like *Diſcharges* of this *Subterranean Fire*, there likewiſe is *mineral Matter*, more or leſs, *hurry'd* up along with it. That even the *Heat* of the *Sun*, and indeed any other, though but an *accidental Heat*, hath the ſame *Effect*, and contributes to the *rai-*

ſing

ſing of *mineral Matter* out of the *Earth.*

That *Ætna*, *Veſuvius*, and the other *Volcanoes*, diſcharge forth, together with the *Fire*, not only *metallick* and *mineral Matter* in great *Quantity*, but *Sand* likewiſe, and huge *Stones*: and with that *Force* too as to toſs them up ſometimes to a very great *Height* in the *Air*.

That the *Heat*, which ariſes out of the *leſſer Spiracles*, alſo brings forth along with it *mineral Matter*, and particularly *Nitre*, and *Sulphur*; ſome of which it affixes to the *Tops* and *Sides* of the *Grotto's* as it paſſes, which *Grotto's* are uſualy ſo *hot* as to ſerve for *natural Stoves*, or *Sweating Vaults*: ſome it *depoſes* near unto, and even upon, the *Surface* of the *Earth*; inſomuch that in ſome Places the *Flores Sulphuris* are gathered in conſiderable *Plenty* near theſe *Spiracles*: ſome it bears in *Steams* up into the *Air*, and this in ſuch *Quantity* too as to be manifeſt to the *Smell*, eſpecialy the *Sulphur*, *that Mineral* ſo particularly affecting *this Senſe*.

That

That the *Heat* which is continualy paſſing up towards the *Thermæ*, brings thither along with it *Particles* of *Spar*, *Alum*, *Sulphur*, *Nitre*, and other *Minerals*, in ſuch *Quantity* that theſe ordinarily as much exceed the common *Acidulæ* * in *Plenty* of this *mineral Matter* as they do in *Heat*. That this *Heat*, aſcending out of the *Thermæ*, *bears up* with it not only *Water*, in Form of *Vapour*, but likewiſe *mineral Matter*; ſome whereof it *affixes* to the Sides and Arches of the *Grotto's* where theſe *Thermæ* ariſe in *ſuch*: or, if they be cover'd with *Buildings*, to the Walls and Roofs of thoſe *Buildings*: to the *Pipes* through which the *Water* is *convey'd*, or the like. That *Sulphur* is in ſome Places collected very *plentifully* adhering to the *Stone* of theſe *Grotts*, and *Buildings*: yea ſometimes *Spar*, and other craſſer *Minerals*, are thus *mounted* up, till, being *ſtop'd* by the *Walls* and *Roofs*, Part *affix* to them, *incruſting* them over, and the reſt are *reverberated* and *form Stalactitæ*, or *Sparry Iceycles* hanging down from the *Arches* of the *Grotto's*, from the *Capitals*

* *Vid. Conſ.* 12. *ſupra.*

Capitals of the *Pillars*, and *Roofs* of the *Buildings*. That where these *Thermæ* are not thus *cover'd* and *vaulted* over, so that the *mineral Matter* is not *stop'd* and *hinder'd* in its *Ascent*, a great Part of it *advances* directly *up* into the *Atmosphere*.

That the *Heat* which is discharged out of the *Earth* at the Time of *Earthquakes* * brings forth *Nitre*, *Sulphur*, and other *mineral Matter* along with it. That the *Water* also which is at the same Time spued out †, thro' the Cracks or Chasmes opened by the *Earthquake*, and thro' the Apertures of *Springs* and *Rivers*, is *turbid* and *stinking*, as being highly saturated with *Sulphureous* and other *mineral Matter*. That the *Acidulæ*, or *Medical Springs* emitt *then* likewise a greater *Quantity* of their *Minerals* than *usual*: and even the *ordinary Springs*, which were before *clear*, *fresh*, and *limpid*, become *thick* and *turbid*, and are impregnated with various *Minerals*, as long as the *Earthquake* lasts. That these *Minerals* do not issue out only at these larger *Exits*, but steam forth likewise thro' the

* *Part* III. *Sect.* 1. *Consf.* 12. *Pag.* 155, & 157. *supra.*

† *Ibid*, *Pag.* 151.

the *Pores* of the *Earth*, occasioning those *Sulphureous*, *Arsenical*, and other *offensive Stenches* which usualy attend *Earthquakes*, and are the *Cause* of *Fevers* and other *malignant Distempers* which commonly *succeed* them; bringing on oftentimes great *Mortalities*, not only amongst *Men*, but even the very *Beasts* and *Fishes*. That these *mineral Eructations* arise in such *Quantity* up into the *Atmosphere* as to *thicken*, *discolour*, and *darken* it sometimes to a very great Degree.

That *any Heat* whatsoever, even an *accidental* one, such as is that which proceeds from the Bodyes of *Animals*, and from their Excrements, promotes the *Ascent* of *mineral Matter*, but more especialy of that which is *subtile*, *light*, and *active*, and is consequently *moveable* more *easily*, and with a *lesser Power*. That by this Means *Nitre* (wherever there happens to be any in the *Earth* underneath) is raised in *Stables*, *Pigeon-Houses*, and other like Receptacles of *Animals*: and in those Places where their Dung lyes heap'd up. That 'twas *this* which occasion'd, ſion'd, in ſome, an *Opinion* that *Nitre* proceeds forth of thoſe *Animals*, and their Excrements; whereas it is found raiſed up, and convened or collected indifferently and as well in *Buildings* where *Animals* rarely or *never come*, as in thoſe they ordinarily frequent; not to mention *that* which is found ſometimes in conſiderable *Plenty* at great *Depths* in the *Earth*: in the *Water* of *Springs*, of *Rivers*, of *Lakes*, and, in ſome *Parts*, even of the *Sea* it ſelf; whereof more largely hereafter. That, in ſuch Places where the *Earth* contains *Nitre* within it, tho' there be no ſuch *adventitious Heat*, if that *Heat* which is almoſt continualy *ſteaming* out of the *Earth* be but *preſerv'd*, its *Diſſipation* prevented, and the *Cold* kept off by ſome Building, or other like Coverture, *this* alone is ordinarily ſufficient to *raiſe* up the *Nitre*, and *bear* it out at the *Surface* of the *Earth*, (unleſs its *Egreſs* be impeded by *Pavements*, or the like *Obſtructions*) and *mount* it up into the *Air*, as far as thoſe *Buildings* will permitt. For, the *Cielings* and *Walls* ſtopping it it in its *Ascent*, it usualy *affixes* unto them, and settles there. And accordingly 'tis frequently found thus *affix'd* to the *Walls* and *Cielings* of *Ground-Rooms*, *Cellars*, and *Vaults*; and this sometimes in such *Quantityes* as to form *nitrose Stalactitæ* *, hanging down from them in Form of *Iceycles*, especialy from the *Tops* and *Arches* of *Cellars* and *Vaults*.

* *Confer* Pag. 222, 223, *supra.*

That the *Heat* of the *Sun* in the *hotter Seasons* being very *intense*, and penetrating the *exteriour* or *superficial* Parts of the *Earth*, it thereby *excites* and *stirs up* those *mineral Exhalations*, in *subterraneous Caverns*, in *Mines*, and in *Cole-pits*, which are commonly called *Damps*. That it is for this Reason that *these* seldom or never happen but in the *Summer Time*; when, the *hotter* the Weather is, the *greater* and *more frequent* are the *Damps*. That, besides *this* of the *Sun*, they are also sometimes *raised* by the Accession of *other Heat*, and particularly by the *Fires* which the *Miners* use in the *Grooves*, for breaking the *Rocks*, and for other Ends. That the *Quantity* of *mineral Matter* thus *raised* is according

cording as there is *more* or *leſs* of it in thoſe *Mines*, eſpecialy of *Sulphur*, *Nitre*, and the like ſubtile and eaſily *moveable* Minerals: and as the *Heat* is *there* more or leſs *intenſe*. That this *mineral Matter* being *ſuſtained* in the *Air* there, and *floating* about in the *Mines*, and *Pits*, it hits upon, and *affixes* it ſelf unto, the *Workmens Tools*, to their *Cloaths*, *Candles*, or any other *Bodyes* that occurr. That where there is any conſiderable *Quantity* of *Sulphur* in the *Exhalation* thus *floating* to and again, it takes *Fire* at the Candles, burns with a *blue Flame*, and emitts a ſtrong *ſulphureous Smell*. That theſe *Damps differ* in their *Effects* according to the *different Minerals* that are the *Cauſe* of them; ours in *England* being generaly reducible to *two Kinds*; whereof one is called the *Suffocating*, the other the *Fulminating Damp*. That the *former* of theſe extinguiſheth the *Candles*, makes the *Workmen* faint, and *vertiginous*, and, when very *great*, *ſuffocates*, and *kills* them. The *Fulminating Damp* will take *Fire* at a *Candle*, or other *Flame*: and, upon its

its *Accenſion*, gives a *Crack* or *Report* like the Diſcharge of a *Gun*, and makes likewiſe an *Exploſion* ſo *forcible* as ſometimes to *kill* the *Miners*, break their *Limbs*, ſhake the *Earth*, and *force Coles*, *Stones*, and other *Bodyes*, even though they be of very *great Weight* and *Bulk*, from the *Bottom* of the *Pit* or *Mine*, up thro' the *Shaft*, diſcharging them out at the *Top* or *Mouth* of it, ſometimes ſtriking off the *Turn* which ſtands thereon, and *mounting* all up to a great *Height* in the *Air*; this being ſucceeded by a *Smoak*, which, both in *Smell*, and all other *Reſpects*, reſembles fired *Gun-powder*: and is, as may appear from theſe and other *Phænomena* of it, nothing but an *Exhalation* of *Nitre* and *Sulphur*, which are the principal *Ingredients* of that *Compoſition* we call *Gunpowder*. That as theſe *Damps* are cauſed by *Heat*, ſo they are remedy'd by withdrawing that *Heat*, and by conveying forth the *mineral Steams*; which the *Miners* effect by *Perflations* with large *Bellows*: by letting down *Tubes*, and ſinking new *Shafts*, whereby they give free *Paſſage* and *Motion*

Motion to the *Air*, which ventilates and cools the *Mines*, purges and frees them from these *mineral Exhalations.*

That at such Time as the *Sun's Power* is so *great* as thus to *penetrate* the *exteriour Parts* of the *Earth* : to *disturb* these *mineral Particles* : and *raise* them from out the *Strata* wherein they lay, it does not only *sustain* them in the *Air* of *Grotto's*, *Mines*, and other *Caverns* under Ground, but likewise *bears* them out thro' the *Mouths* of those *Caverns*, and thro' the ordinary *Cracks* and *Pores* of the *Earth*, *mounting* them up, along with the *watery Exhalations*, into the *Atmosphere* *, especialy *Sulphur*, *Nitre*, and the other more *light* and *active Minerals* ; where they form *Meteors* : and are particularly the Cause of *Thunder*, and of *Lightning*. That, this *mineral Matter* requiring a considerable Degree of *Heat* to *raise* it, the most *Northern Climes*, and the *Winter Seasons*, are, for that Reason, little or not at all troubled with *Thunder* ; it seldom happening, in any great Measure, but in the *hotter Months*,

* *Vid.* *Part* III. *Sect.* 1. *Consect.* 8.

and in the *Southern Countries*, as in *Congo*, *Guinea*, and other Parts of *Africa*, and in the *Southern Parts* of *Asia* and *America*; where 'tis, during the *Season* of their *great Rains**, horribly *loud* and *astonishing*, and as much exceeds the *Thunder* of these *Northern Climes*, as the *Heat there* exceeds that of *these Climes*. That the *mineral Matter* which is *discharged* forth of *Volcano's*, and other like *Spiracles*: and out of the *Thermæ*, ascends up into the *Air*, and contributes to the *Formation* of these *Meteors*. That likewise the *Nitre* and *Sulphur*, which are belch'd forth of the *Earth* at the Time of *Earthquakes* (the *Countryes* which are most obnoxious to this *Malady* abounding, as I have already intimated*, with these two *Minerals* particularly) in such *Plènty* as to thicken and darken the *Air*, constitute there a kind of *Aerial Gunpowder*, and are the *Cause* of that dismal and terrible *Thunder* and *Lightning* which commonly, if not always, attend *Earthquakes*; even when all was till then *calm* and *clear*, and there was not the least *Sign* or *Presage*

* *Vid. Pag.* 141. *supra.*

* *Vid. Pag.* 157. *supra.*

Preſage of any ſuch Thing before the *Earthquake* began.

That as the *mineral Eruptions* which happen during *Earthquakes**: and the *Steams* and *Damps* of *Mines*†, are detrimental to *Health*, hurtful and injurious to the Bodyes of *Men* and other *Animals*, ſo likewiſe are the *mineral Exhalations* which are thus *raiſed* by the *Sun* out of the Body of the *Earth* up into the *Atmoſphere*; but more eſpecialy in thoſe *Parts* of it where there are *Arſenical*, or other like *noxious Minerals* lodg'd underneath. That *theſe* mingling with, and being diſſeminated in the *Air*, and paſſing together with it into the *Lungs* in *Reſpiration*, are by them tranſmitted into the *Body*, where they *infect* the Maſs of *Blood*, create *Perturbations* and diſorderly *Motions* therein, and lay the Foundation of *Peſtilential Fevers*, and other malignant *Diſtempers*. That 'tis for this Reaſon that the *Southern Countryes* are more frequently moleſted and incommoded by theſe *Diſtempers* than the *Northern* are: and that they are more rife and ſtirring in the *hotter Months*, in *July*

* *Vid.* Pag. 224. *ſupra.*

† *Vid.* Pag. 227, 228, *ſupra.*

and *August*, than in the *colder*, *December*, *January*, and the reſt. That 'tis indeed true, that in *September* and *October*, which are none of the *hotteſt Months*, theſe *Diſeaſes* are oftentimes as *epidemical* as in the precedent and *warmer Seaſon:* and do not *abate* and *remitt* in Proportion to the *Remiſſion* of the *Sun's Heat* in thoſe Months. But this is purely *accidental*, and happens meerly becauſe the *Heat within* the *Earth* is not liable to ſo *ſudden Viciſſitudes*, or ſo *quickly ſpent* and *diſpers'd*, as is that which is *upon* it, and in the *Air*. *This* therefore, the *Pores* of the *Earth* remaining ſtill as *free*, and *open*, as ever, continues to *ſend out* the *mineral Steams* as before, but in *leſſer* and *leſſer Quantity*, anſwerably to the gradual *Diminution* of this *Heat*. Which *Steams*, though now ſent up to the *Surface* of the *Earth* only in *leſſer Plenty*, may be much more *offenſive* and *miſchievous* than in the *hotter Months* when they came forth in far *greater*. For the *Sun's* Power being in *thoſe Months* alſo *greater*, it then ſtraitways *hurryes* theſe *Steams* up into the *Atmoſphere*.

ſphere. Whereas in the *colder*, its Power being *leſſen'd*, it cannot *bear* them up ſo *faſt*; ſo that they then *ſtay* and *ſtagnate* near the *Surface* of the *Earth*, *ſwimming* and *floating a-bout* in *that Region* of the *Air* where-in *we breath*; where they muſt needs be much more *pernicious* than when borne up to a *greater Height*, and ſo farther out of the Way. And this is indeed much the *Caſe* of *Foggs*; particularly of thoſe which we frequently obſerve after *Sun-ſetting*, even in our *hotteſt Months*. Theſe are nothing but a *Vapour*, conſiſting of *Water*, and of ſuch *mineral Matter* as this met with in its Paſſage, and could well *bring up* along with it. Which *Vapour* was ſent up in *greater Quantity* all the *foregoing Day*, than now in the *E-vening*. But the Sun, *then* being above the *Horizon*, taking it at the *Surface* of the *Earth*, and rapidly *mounting* it up into the *Atmoſphere*, it was not *diſcernible*, as *now* it is; becauſe, the *Sun* being now *gone off*, and ceaſing any longer to *operate* upon it, the *Vapour ſtagnates at* and *near* the *Earth*, and ſaturates the

Air till 'tis ſo *thick* as to be eaſily *viſible* therein. And when at length the *Heat* there is ſomewhat *further ſpent*, which is uſualy about the *Middle* of the *Night*, it *falls* down again in a *Dew*, alighting upon *Herbs* and other *Vegetables*, which it *cheriſhes*, *cools*, and *refreſhes*, after the *ſcorching Heat* of the *foregoing Day*. But if it happens, as ſometimes it does, that *this Vapour* bears up along with it any *noxious mineral Steams*, it then *blaſts Vegetables*, eſpecialy thoſe which are more *young* and *tender*; *blights Corn*, and *Fruits*: and is ſometimes *injurious* even to *Men* who chance to be then *abroad* in the *Fields*. 'Tis alſo the *Caſe* of *Water* at the *Surface* of the *Earth**; where the *Springs* and *Rivers* are very *low*, yea ſome of them *ceaſe* to yield any *Water* at all, in the *Summer Months*; becauſe the *Sun's Power* is *then* ſo *great* as eaſily and ſpeedily to *bear up* into the *Atmoſphere*, in ſmall and inviſible Parcels, and in Form of an extremely fine and thin *Vapour*, a very great Part of the *Water* which is *ſent up* out of the *Abyſs*. Whereas, in the *Winter-time*,

* *Confer* *Part* III. *Sect.* 1. *Conſect.* 8.

time, the *Sun* is withdrawn *farther off*, and its *Power leſſen'd*, ſo that it cannot *then* buoy it up as before; for which Reaſon 'tis that ſo much more of it *then* ſtands at the *Surface* of the *Earth*, and *ſtagnates* there. So likewiſe for *Rain*; we learn from *Experiment* that there commonly falls in *England*, in *France*, and ſome other Countryes, more Rain in *June* and *July*, than in *December* and *January*. But it makes a much *greater Shew* upon the Earth in *theſe Months* than in *thoſe*, becauſe it lyes *longer* upon it; the *Sun* now wanting *Power* to *exhale* and bear it up ſo *quickly* and *plentifully* as *then* it did. 'Tis likewiſe the *Caſe* of the *Halitus* emitted forth of the *Lungs* of *Men* and other *Animals*. In a *Phyſiological Treatiſe*, which I have by me, concerning the *Structure and Uſe of the Parts of Animals*, diſcourſing of the *Lungs*, I ſhew that they are one *grand Emunctory* of the *Body*: that the *main End* of *Reſpiration* is continualy to *diſcharge* and *expell* an *excrementitious Fluid* out of the *Maſs* of *Blood*: and I prove from ſeveral *Experiments* that there passes out of the *Body* a *greater Quantity* of *Fluid Matter* this Way, I mean upwards, and through the *Lungs*, than there does of *Urine*, by the *Kidneys*, downwards. Now the *Fluid*, which is thus *secreted*, and *expired* forth along with the *Air*, goes off with it in insensible Parcels, in the *Summer Season*, when the *ambient Air* contains *Heat* enough to *bear* it quickly away, and so *disperse* it. But, in the *Winter*, when the *Heat* without is *less*, it oftentimes becomes so far *condens'd* as to be *visible*, flowing out of the *Mouth* in Form of a *Fume*, or *crasser Vapour*: and may, by proper *Vessels*, set in a strong *freezing Mixture*, the better to *condense* this *Vapour*, be *collected* in *considerable Quantity*. But to return. That 'tis not without a very extraordinary *Providence* that there so constantly happens, in the Month of *September* (the *Time* when chiefly these *mineral Steams* stagnate thus at and near the Surface of the Earth) a very *nipping* and *severe Season* of *Cold*, far beyond what might, from the *Sun's Height* and *Power*, be *then* expected:

pected: beyond that of *October* and *November*: and ſometimes equal to that of *January*, and the *coldeſt Months*: as alſo that there then ſo conſtantly happens very *bluſtering* and *turbulent Winds*; the *Cold* ſerving to *check* and put a *Stop* to the *Aſcent* of this *mineral Matter*: and the *Winds* to *diſſipate* and *convey away* that which was before *rais'd* out of the *Earth*; which, was it not thus *carry'd off*, would be infinitely more *fatal* and *pernicious* to *Man* and other *Animals* than *now* it is. But I muſt be contented here to give only ſhort *Hints* of theſe, as of other Things: and to write but obſcurely and reſervedly, untill I have Opportunity to expreſs my *Sentiments* of them with greater Copiouſneſs, Freedom, and Perſpicuity.

Thus much of the *Scheme* of my *Deſign* in this Part have I run over: and led my Reader a long and tedious Jaunt in tracing out theſe *metallick* and *mineral Bodyes*: in purſuing them through their ſeveral *Mazes* and *Retreats*: through the *Earth*, the *Water*, and the *Air*. And yet,

yet, long as it is, we are not much farther than the *Borders* of the *mineral Kingdom*, and have done little more yet than ſettled and adjuſted *Preliminaries*; ſo very *ample* is this *Kingdom*, ſo *various* and *manifold* its *Productions*. For the foregoing *Concluſions* relate only to the *Origin* and *Growth* of *theſe Bodyes*; the *Natural Hiſtory* of each *particular Metall* and *Mineral*, with the *Obſervations* whereon that *Hiſtory* is grounded, being ſtill to come. But I muſt be forced wholey to wave and ſuperſede the Detail of theſe; for I perceive, do what I can, this *Abſtract* will ſwell much beyond the Bounds which I at firſt deſign'd.

This Fourth Part will be follow'd by ſeveral *Treatiſes*, ſerving to confirm, and to illuſtrate ſome *Paſſages* in it; whereof I ſhall at preſent only mention the *four* following.

1. *Rules* and *Directions* for the *Diſcovery* of *Metalls* and *Minerals* latent in the *Earth*; with an *Enquiry* why theſe lye ſometimes ſo near the *Surface*, and did not, (becauſe of their *greater Gravity*) at the *General Subſidence* in the *Deluge**, fall

* *Part* II. *Conſect.* 3. & *Pt.* IV. *Conſect.* 3.

fall to a much *greater Depth* than we now find them: even to ſuch a *Depth* as to have lain quite out of *humane Reach*, and ſo have been all *bury'd*, and irrecoverably *loſt*.

2. An examination of the Common Doctrine about the *Generation* of *Metalls* and *Minerals:* and particularly *that* of the *Chymiſts*; with an Appendix, relating to the *Tranſmutation* of *Metalls*; detecting the Impoſtures and Eluſions of thoſe who have pretended to it: and evincing the *Impoſſibility* of it from the moſt plain, ſimple, and Phyſical *Reaſons:* proving likewiſe that there are no ſuch natural *Gradations*, and *Converſions* of one *Metall* and *Mineral* into another, in the *Earth*, as many have fancy'd. As alſo an Account of the *mineral Juyces* in the *Earth*, which ſome *Writers* have imagined to be I know not what *Seeds* of *Minerals*; ſhewing that they are, for the far greateſt Part, nothing but *Water* ſtrongly impregnated with *mineral Matter*, which it derives from the *Strata* as it paſſes through them*.

3. Rela-

* *Conſ.* 5. *&c. ſupra.*

3. Relations, obtain'd from ſeveral Hands, concerning the *State* of *Metalls* and *Minerals* in *Foreign Countries*: in divers Parts of *Aſia*, *Africa*, and *America*, as well as in *Hungary*, *Germany*, *Sweden*, and other Parts of *Europe*: and particularly of thoſe which are not found in *England*, ſhewing that the *Condition* of theſe *Bodyes* in thoſe remoter *Regions* is exactly conformable to that of *ours* here: and that they were all put into this *Condition* by the very *ſame Means* *.

* *Confer Conſect.* I. *ſupra.*

4. *Obſervations* concerning Engliſh *Amber*: and *Relations*, from abroad, about the *Amber* of *Pruſſia*, and other diſtant Places. With a *Diſcourſe*, founded upon them, proving that Amber is not a *gummous* or *reſinous Subſtance* drawn out of Trees by the Sun's Heat, and coagulated and harden'd by falling down into Rivers, or the Sea, as the *Ancients* generaly believ'd: but is a *Natural Foſſil*, as Pebles, Flints, *Pyritæ*, and the like, are: form'd at the *ſame Time*, and by the *ſame Means* that they were: and all of it originaly repos'd in the *Strata* of Earth,

Earth, Sand, *&c.* together with them. That it is indeed found in ſome Places lying upon the *Shores* of the *Sea*, and of *Rivers*; but 'tis alſo found at *Land*, and dug up (ſometimes at very great *Depths*) in the *Earth:* and this as well in Places very *remote* from any *Sea*, or *River*, as in thoſe which are nearer unto them. That 'tis *digg'd* out of even the *higheſt Mountains*, and indeed all other *Parts* of the *Earth*, contingently, and indifferently, as the *Pyritæ*, Agates, Jaſpers, Pebles, and the reſt, are. That wherever 'tis found upon the *Sea-Shores*, there alſo is it as certainly found at *Land*, up in the neighbouring Country: and particularly in *Pruſſia*, upon whoſe *Shores* ſo great a *Quantity* of *Amber* is yearly collected, 'tis dig'd up almoſt all over the *Country*. That even that which now lyes looſe upon the *Sea-Shores*, was all of it *originaly* lodg'd in the *Earth:* in the *Strata* of Sand, Marle, Clay, and the like, whereof the neighbouring *Land*, and the *Cliffs* adjacent to thoſe *Shores*, do conſiſt; and wherever 'tis ſo found ſcatter'd upon the *Shores*,

Shores, there is it as constantly found lodg'd in the *Cliffs* thereabouts. That when the *Sea*, at High-water, comes up unto, and bears hard upon the said *Cliffs*, and is agitated by *Winds* and *Storms*, it frequently *beats down* huge Pieces of *Earth* from them. Which *Earth*, falling into the Water, is, by its continued *Agitation* and *Motion* dissolved : and borne by Degrees down into the *Sea*, being loose, and light, and so easily reduced into lesser Parcels, dissipated, and wash'd away. But the Pebles, *Pyritæ*, *Amber*, or other like *Nodules*, which happen'd to be repos'd in those *Cliffs*, amongst the *Earth* so beaten down, being *hard*, and not so dissoluble, and likewise more *bulky* and *ponderous*, are left behind upon the *Shores*; being impeded, and secur'd, by that their *Bulk* and *Weight*, from being born, along with the *Terrestrial Matter*, into the *Sea*. That therefore the *Sea* is no ways concern'd in the *Formation* of these *Bodies* : no more in the *Formation* of *Amber*, than of the *Pyritæ*, Flints, and other *mineral Masses* that are found together

ther with it: but only *diſlodges* and *diſcovers* them, bears away the *Earth* wherein they were *bury'd*, waſhes off the *Soil* and *Sordes* wherein they were *involv'd* and *conceal'd*, and thereby renders them more *conſpicuous*, *apparent*, and eaſy to be *found*. That this is ſo known and experienc'd amongſt the *People*, who are employ'd to *gather* the *Amber*, that they always run down to the Sea Side, for that Purpoſe, after a *Storm*: and, if it hath been ſo *great* as to beat down Part of the *Cliffs* there, they aſſuredly find *Amber*, more or leſs, upon the *Seas Ebb* and Retirement, and after every *Retreat* of the *Sea* for ſome *Tides* after; the *Sea* not bearing down the *Earth* immediately and all at once, but waſhing it off by little and little, and ſo *diſcovering* the *Amber* by Degrees, ſome after one *Tide*, and ſome after another. That particularly the *Amber*, *Vitriolick Pyritæ*, and other like *Bodyes*, that are found upon the *Shores* of *Kent*, *Eſſex*, *Hampſhire*, and elſewhere, all came firſt of the *bordering Cliffs*, and were diſlodged by *this Means*: and

are

are found in the *Earth*, as well as upon the *Shores*, whenever 'tis laid open, as in ſinking *Wells*, *Pits*, and the like. That not only the *Sea*, but *Rivers* and *Rains* alſo, are inſtrumental to the *Diſcovering* of *Amber*, and other *Foſſils*, by waſhing away the Earth and Dirt that before cover'd and concealed them. Thus the *Golden Nodules*, or, as they are commonly call'd, *Gold-grains*, *Amethiſtine Pebles*, *Amber*, and other *Stones* of *Worth*, are uncover'd by ſuch *Rivers* as chance to run through the *Grounds* which *contain* thoſe *Bodyes* in them. Thus likewiſe *Rains*, by their *waſhing* the *Earth* down from off the *Hills* *, clear, and diſcloſe ſuch *Pyritæ*, *Selenitæ*, or other *Bodyes*, that happen to be lodg'd, near the *Surface* of the *Earth*, in thoſe *Hills*. And 'tis by this Means chiefly that the *Grain-Gold*, upon all the *Gold Coaſt* (as 'tis call'd) in *Guinea*, is diſplay'd; the *Rains*, falling there in great *Abundance*, and with incredible *Force*, thereby the more powerfully *beating off* the *Earth*. This the *Negroes*, Natives of thoſe Parts, know full well:

* *Part* V. *Conſect.* 2.

well: and therefore do not expect to find much of it unless after the *Season* of their *Rains* *; when they never fail to find it, no more than the *Amber Gatherers* fail of finding *that* upon the *Sea-Coasts* after a *Storm*. And if those *Persons* who are curious in *collecting* either *Minerals*, or the *Shells*, *Teeth*, or other *Parts* of *Animal Bodyes*, that have been *buried* in the *Earth*, do but search the *Hills* after *Rains*, and the *Sea-Shores* after *Storms*, I dare undertake they will not lose their Labour. But to return. That *Amber* is not only lodged in the *Strata* of *Earth*, and of *Sand*, together with the other *mineral Nodules*, but is sometimes found actualy growing unto, and combined into the same *Mass* * with the *Pyrites*, and *others* of them. That it likewise contains in it *Insects*, *Flies*, *Shells*, and other *heterogeneous Bodyes*, in like Manner as the *Pyritæ*, *Flints*, and all other analogous *Fossils* do *. That altho' *Amber* be most commonly of a *yellowish Colour*, and therefore not unlike some Kinds of *Gumms*, yet there is found of it also of several

* *Vid.* Part III. *Sect.* 1. *Consect.* 8. *Pag.* 141.

* *Vid. Consect.* 2. *supra.*

* *Ibid.*

other *Colours*, as *Black*, *White*, *Brown*, *Green*, *Blue*, and *Purple*, to name no more. Yea the very ſame *Lump* is frequently of *different Colours*. That theſe *Colours* are all *accidental*, even the *yellow* it ſelf, and owing to the *Intermixture* of *foreign Matter*, which concreted into the ſame *Maſs* with the *proper Matter* of this *Stone*, and with the *heterogeneous Bodyes* which are *included* in it, at the Time of its *Coalition**. That this is the Caſe of *Agates*, of *Cornelians*, of *Topazes*, and many other *coloured Stones*; the *Colours* of ſeveral whereof, and even *that* of *Amber* it ſelf, may, by a very eaſy *Proceſs*, be, in great Meaſure, if not wholey, *extracted*, and taken from them: and the *Bodyes* of theſe *Stones* rendred almoſt, if not quite, as *pellucid* as *Cryſtall*, without ſenſibly damaging the *Texture* of them. That even the moſt *obvious* and *ordinary Minerals* are not free from this Contagion of *adventitious Matter*; *Common Salt* it ſelf, when found naturaly *cryſtallized* amongſt other *Minerals* and *Metalls*, in the *perpendicular Intervalls* of the *Strata* of Stone,

* *Ibid.*

Stone, being, not only *pellucid*, as it naturaly is when *pure* and *ſimple*, put *white* alſo, and like the *white cryſtallized Spar* : *yellow*, and nearly reſembling the *Topaz* : *blue*, and not unlike the *Saphire* ; and yet theſe ſpecious *Bodyes*, and *Gemms* as to outward Shew, upon *Tryal*, yield nothing but *meer Salt*, with an extremely ſmall *Admixture* of other *Matter*, which gave them their *Tincture*. Which may ſerve for a further *Inſtance* of the *confuſed State* of *Minerals* in the *Earth* : and of the *Uncertainty* of their *Colours*, and *Figures* *.

* *Confer pag.* 190, *to* 193.

PART V.

Of the Alterations *which the* Terraqueous Globe *hath undergone* since *the Time of the* Deluge.

IT now remains that we take a View of the *Postdiluvian State* of this our *Globe*: that we examine how it hath *stood* for this last four thousand Years: that we enquire what *Accidents* have befallen it, and what *Alterations* it hath suffered *since* that wonderful *Change* it underwent at the *Deluge*.

There have been some who have made a mighty Outcry about *Changes* and *Alterations* in the *Terraqueous Globe*. The *Pretences* and *Pleas* of each I consider in the *first Part* of this *Essay*; shewing that they are without

without any juſt Grounds: and that there are no *Signs*, or *Footſteps*, in the whole *Globe*, of ſuch *Alterations*. And indeed 'tis well for the World that there are not; for the *Alterations*, which they have fancy'd, are ſuch as turn all the *wrong Way*: ſuch as are without *Uſe*, and have no *End* at all, or, which is worſe than none, a *bad* one: and tend to the *Damage* and *Detriment* of the *Earth* and its *Productions*.

Notwithſtanding, ſome *Alterations* there are which it hath, and doth ſtill undergo. This is what we may pronounce with *Certainty*: and there want not *Inſtances* enough ſufficiently to vouch and atteſt it. But theſe *Alterations* are of a quite different *Strain*: theſe are as *amicable* and *beneficent* to the *Earth* and *Terreſtrial Bodyes*, as the *others*, were there realy ſuch, would be *pernicious* and *deſtructive* to both. I have already given* ſome *Intimations* of the *Changes* which happen in the *interiour Parts* of the *Earth*, I mean the *Tranſitions* and *Removes* of *Metalls* and *Minerals* there: and ſhewn of

* *Part* IV. *Conſect.* 4.

what *Use* and *Advantage* those *Changes* are to the *World**. So that I may now pass freely on to consider *those* which befall the *exteriour*, or *Surface* of it. And *these* are brought about silently and insensibly: and, which is the constant Method of *Nature*, with all imaginable *Benignity* and *Gentleness*. Here is none of the *Hurry* and *Precipitation*, none of the *Blustering* and *Violence*: no more than any of the direful and ruinous *Effects*, which must needs have attended those *Supposititious Changes*. And as these *Alterations* are not *great*, so neither are they *numerous*. I have made careful Search *on all Hands*, and canvass'd the Matter with all possible *Diligence*: and yet there are none that I can advance from my own *Observations*, but,

* *Ibid. Consect.* 9.

1. That the *upper* or outermost *Stratum* of *Earth*, that *Stratum* whereon Men and other Animals tread, and Vegetables grow, is in a *perpetual Flux*, and *Change*; this being the *common Fund* and *Promptuary* that supplyes and sends forth *Matter* for the *Formation* of *Bodyes* upon the *Face* of the *Earth*. That all

Animals,

Animals, and particularly *Mankind*, as well as *all Vegetables*, which have had *Being* ſince the *Creation* of the *World*, derived all the *conſtituent Matter* of their *Bodyes* ſucceſſively, in *all Ages*, out of this *Fund*.

That the *Matter*, which is thus *drawn* out of this *Stratum* for the *Formation* of theſe *Bodies*, is at length laid down again in it, and *reſtored back* unto it, upon the *Diſſolution* of them; where it lies ready to be *again aſſumed*, and *educed* thence for the *fitting forth* of other *like Bodyes*, in a *continual Succeſſion*.

That the *conſtituent Matter* of any one *Body* being *proper*, and turning thus naturaly, when again *refunded* into this *Stratum*, to the *Conſtitution* of another *like Body*, there is a kind of *Revolution* or *Circulation* of it; ſo that the *Stock* or *Fund* can never poſſibly be *exhauſted*, nor the *Flux* and *Alteration* ſenſible.

That as the *Bodyes* which ariſe out of this *Fund* are *various*, differing very much, not only from one another, but the *Members*, *Organs*, or *Parts* of each individual amongſt themſelves; ſo likewiſe is the *Mat-*

ter of this *Fund* whereof they all *consist*. For though, when confusedly *blended* and *mingled*, as it is whilst lying in this *Stratum*, it may put on a *Face* never so uniform and alike, yet it is in reality very *different*, and consists of *several Ranks*, *Sets*, or *Kinds* of *Corpuscles*.

That all the *Corpuscles* that are of the *same Set*, or *Kind*, agree in every Thing, and are most exactly *like* unto each other in all *Respects*. But those that are of *diverse Kinds*, differ from one another, as well in *Matter* or *Substance*, in *Specifick Gravity*, in *Hardness*, in *Flexibility*, and several *other Ways*, as in *Bigness*, and *Figure*. That from the various *Composures* and *Combinations* of these *Corpuscles* together, happen all the *Varieties* of the *Bodyes* formed out of them: all their *Differences* in *Colour* and outward *Appearance*, in *Taste*, in *Smell*, in *Hardness*, in *specifick Gravity*, and all other Regards; in much the same manner as that vast Variety we see of *Words* arises from the various *Order* and *Composition* of the twenty four *Letters* of the Alphabet. But of *this Matter*, and

and of the *Proceſs* and *Method* of *Nature* in the *Formation* of *Bodyes* out of it, I ſhall treat more at large in the *Diſcourſe* it ſelf; wherein I ſhall alſo conſider the *Opinions* of *Others*, particularly the *Ancients*, and, amongſt the reſt, of *Thales* and *Pythagoras*, about the *Elements* or *Principles* of *Natural Things*; for I now haſten to propoſe the *other Alterations* that happen in the *Terraqueous Globe*.

That *Rocks*, *Mountains*, and the 2.
other *Elevations* of the *Earth* (eſpecially thoſe whoſe *Surfaces* are yearly ſtirr'd and diſturbed by digging, plowing, or the like) ſuffer a continual *Decrement*, and grow *lower* and *lower*; the ſuperficial Parts of them being by little and little waſh'd away by *Rains*, and *borne down* upon the ſubjacent *Plains* and *Valleys*. That even the *Stone* it ſelf (whether naked and uncover'd as in *Rocks*, or inveſted with a *Stratum* of Earth as is that in our ordinary *Hills*) is not, by its Solidity, privileged and ſecur'd againſt them, but is *diſſolved* by Degrees, and waſh'd alſo down, in

in its. turn, as well as the looſer Earth.

3. That the *Matter*, which thus devolves from the *Hills* down upon the *lower Grounds*, does not conſiderably *raiſe* and *augment* them; a great *Part of it*, *viz.* the *vegetative* and *lighter Terreſtrial Matter*, being either mounted up into the *Atmoſphere* by the aſcending Vapour*, or carry'd along with the Rain-water into *Rivers*, and, by them, into the *Sea* †; whence 'tis returned back again to the Earth diſperſedly by *Rain* †, and ſerves for the Nutriment and Formation of the *Plants* which grow thereon: and *the reſt* of it, being more craſs and ponderous, does not *move far*, but lodges in the Clefts, Craggs, and *Sides* of the Rocks or Mountains, and at or near the *Roots* or *Bottoms* of them.

* *Vid.* *Part* III. *Sect.* 1. *Conſect.* 8.
† † *Confer* *Pag.* 50 & *ſeq.* *uti* & *Pag.* 143. & *ſeqq.*

4. That the *Stone* of *Rocks* and *Mountains* being by Degrees in this Manner *diſſolved*, and the *Sand* borne off, the *Shells*, and other *Marine Bodyes* which were originaly *included* therein*, are by that Means *let looſe*, turned out, and expoſed upon the *Surface* of the *Earth*. That 'tis for this

* *Part* II. *Conſect.* 3.

this Reaſon that theſe *Marine Bodyes* are now moſt commonly found upon *Hills*, and the *higher Grounds*; thoſe *few* which are found *below* and at the *Bottoms* of them, being for the moſt part only ſuch as have *fallen down* from above, and from the *Tops* of them. For thoſe which were, at the Time of the *Deluge*, repoſed upon the *Surface* of the *Earth*, are moſt of them *periſh'd* and *gone**. And indeed *theſe*, which are yet exiſtent, are *preſerv'd* only accidentaly, by their being at firſt *encloſed* in the *Strata* of Stone, or other cloſe Matter, and ſo ſecured by it as long as it was it ſelf ſecure, I mean, until it was thus *diſſolv'd*, and ſo could no longer contribute any thing to their *Preſervation*.

* *Vid.* *Pag.* 70, & 71, *ſupra.*

5\. That theſe *Shells* and other *Bodyes*, being thus *turned out* of the *Stone*, and *expoſed* looſe upon the *Surface* of the *Earth*, to the Injuries of *Weather*, and of the *Plough*, to be *trod upon* by Horſes and other Cattle, and to many other *external Accidents*, are, in Tract of Time, *worn*, *fretted*, and *broken* to Pieces.

That

That the *Shells* being ſo broken, ſtruck off, and gone, the *Stone included* in them is thereby *diſcloſed* and ſet at Liberty; which *Stone* conſiſts of the *Sand* wherewith the *Cavityes* of thoſe *Shells* were filled when they were *ſuſtained* together with it in the *Water* at the *Deluge* *, and which at length *ſubſided* in them, and was *lodged* with them in the *Strata* of Sand-Stone; the *Sand* contained within the *Shell* becoming *ſolid* and *conſiſtent* at the ſame Time that the *ambient*, or that of the *Stratum* without, did *.

* *Part* II. *Conſ.*2.& 3

* *Part* II. *Conſect.* 4.

That therefore the *Shells* ſerved as *Plaſms* or *Moulds* to this *Sand*; which, when *conſolidated*, and afterwards in tract of Time by this means *freed* from its *inveſtient Shell*, is of the ſame *Shape* and *Size* as is the *Cavity* of the *Shell*, of what Kind ſoever that *Shell* happened to be. That this is the true *Origin* of thoſe *Stones* (conſiſting of *Sand* *) which are called, by Authors, *Cochlitæ*, *Conchitæ*, *Muitæ*, *Oſtracitæ*, *Ctenitæ* †, *&c.* and which are of *conſtant*, *regular*, and *ſpecifick Figures*; as are the *Cochleæ*, *Conchæ*, and the other

* *Thoſe which conſiſt of* Spar, Flint, *&c. I have conſider'd above*, Part 4. *Conſ.* 2.

† *Vulgarly* Pectinitæ.

other *Shells* in which they were *moulded*, and from which, by reason of their ſo near *Reſemblance* of the Inſides of them, they borrow their ſeveral *Denominations*.

That theſe *formed Stones* being by this Means deſpoil'd of their *Shells*, and expoſed naked, upon the Surface of the Ground, to the *Injuries* before recited, do alſo *themſelves* in Time *decay*, *wear*, and *moulder* away, and are frequently found *defaced* and *broken* to Pieces; in like Manner as the *Strata* of Stone, wherein they were originaly lodged, firſt did: and afterwards the *Shells*, wherein theſe *Stones* were encloſed and formed. 6.

This *Deterration*, as 'tis call'd, or Devolution of Earth and Sand from the *Mountains* and *higher Grounds*, is not in equal *Quantity* and alike in all Places, but *varies* according to the different *Height* of thoſe *Mountains*, and to the *Extent* of the *Plane* at Top of them: to the different *Conſiſtence* and Durableneſs of the *Strata* whereof they conſiſt: and according as they are more or leſs diſturbed by *Showers* *, *Plowing*, or other

* *Which are greater, and fall with more Violence in ſome Countries than in others.* Vid. Part 3. Conſ. 8.

other *Accidents*. Nay this *Deterration* varies in different Parts of even the *ſame Mountain*; thoſe which lye nearer to the *Brink* or *Margin* of it ſuffering a quicker and greater *Decrement* than thoſe which are more remote therefrom, and towards the *Middle* of it. But, though this Devolution be thus different, 'tis not any where, even where greateſt, very *conſiderable*: and therefore does not make any great *Alteration* in the Face of the Earth. This I have learn'd from *Obſervations* purpoſely made in ſeveral Parts of *England*: and when I ſhall, in the larger Work, propoſe the *Standard* whereby I give *Judgment* of it, any one may preſently and eaſily inform himſelf of the *Quantity* and *Meaſure* of it, either *here*, or in any other Part of the *World*.

There are indeed ſome other *Caſualties* that the *Globe* is obnoxious unto, ſuch as *Earthquakes*, and the burning Mountains, or *Volcanoes*. But of *theſe*, I thank God, and the good Conſtitution of this happy Iſland, I have not had an Opportunity of *Obſervation*. Yet ſomething

I have

I have to offer concerning *theſe*, and the *Cauſes* of them, from the *Obſervations* of *others*. Not that the Thing is ſo very material, or that they make ſuch Havock, and *Alterations* in the *Globe* as ſome Men fancy. We have Aſſurance from *Hiſtory*, that *Ætna* and *Veſuvius* have ſent forth Flames, by fits, for theſe two or three thouſand Years: and no doubt but they have done ſo much longer. And yet we ſee both *Sicily* and *Campania*, the Countries wherein thoſe two Mountains ſtand, are ſtill where they were: nay the very Mountains themſelves are yet in *Being*, and have not ſuffered any conſiderable *Diminution* or *Conſumption*, but are at this Day the two *higheſt* Mountains in thoſe Countries. What they have realy ſuffer'd: by what Means both *theſe*, and *Earthquakes*, are occaſion'd: and what are their *Effects* upon the *Globe*, ſhall be fully and carefully conſider'd in due Place. From which Conſiderations it will appear, that even *theſe* have their *Uſes*: and that, although they do make ſome leſſer *Alterations* in ſome few Parts of the *Earth*,

Earth, and sometimes molest and incommode the Inhabitants of those Parts, yet the *Agent*, whereby both the one and the other is effected, is of that indispensible *Necessity* and *Use* both to the *Earth* it self, to *Mankind*, and to all other the *Productions* of it, that they could not *subsist* without it. I have already given some brief Intimations that *Winds* and *Hurricanes* at Land*, *Tempests* and *Storms* at Sea†, (Things that have always been look'd upon with as evil an Eye as Earthquakes, and pointed at as only disastrous and mischievous to the World) are yet not without a very necessary and excellent *Use*: the same have I also done concerning *Volcanoes**; but I must not dwell upon these Things too long, wherefore I shall only now dispatch what is further necessary to be hinted here about this *Decrement* of *Mountains*, and then conclude this Part.

* *Part* IV. *Cons.* 14. *Pag.* 237.

† *Vid. Pag.* 51, 52, *&c.*

* *Part* III. *Sect.* 1. *Cons.* 13.

And *this*, as it does not make any great *Alteration*, so neither doth that, which it realy does make, any ways *endamage* or disorder the *Globe*. 'Tis not any the least *Detriment* or *Disadvantage*

Disadvantage to the *Productions* of it, to Vegetables, to Animals, and particularly to Mankind. Nor does it thwart and interfere with the grand *Design* of *Providence*, *viz.* the *Conservation* of the *Globe*, and the *Propagation* of *Bodies* upon it, for the *Use* of *Man.* So far from this, that it is very highly *beneficial* and *serviceable* to *both*; which will farther appear if we consider,

That in the *first Ages* after the *Deluge*, when the *Number* of *Mankind*, of *Quadrupeds*, and of the other *Animals* was but *small*, the *Valleys* and *Plains* were more than sufficient for their *Habitation* and *Use.* And, by such Time as that Stock of them was *inlarged*, that they were further spread and *multiplyed*, and thereby the Earth so far peopled and replenish'd that the *Hills* and higher Grounds began to be *needed*, those *Rocks* and *Mountains*, which in the *first Ages* were *high*, *steep*, and *craggy*, and consequently *then* inconvenient and unfit for *Habitation*, were, by this continual *Deterration*, brought to a *lower Pitch*, render'd more *plain* and *even*, and reduced nearer to the *ordinary Level* of the *Earth*. By which Means they were made *habitable*, by ſuch Time as there was *Occaſion* for them : and *fit* for *Tillage*, for the Production of *Vegetables*, of Corn, and other *Neceſſaries*, for the Uſe of their Inhabitants.

That although the *principal Intention* in the *Precipitation* of the *Vegetative Terreſtrial Matter* *, at the *Deluge*, and the *burying* it in the *Strata* underneath, amongſt the Sand, and other mineral Matter, was to *retrench* and abridge the *Luxury* and *Superabundance* of the Productions of the *Earth*, which had been ſo ingratefully and ſcandalouſly abuſed by its former *Inhabitants*, and to cauſe it to deal them forth for the future more frugaly and *ſparingly*, yet there was a ſtill *further Deſign* in that *Precipitation* : and the Matter, ſo *bury'd*, was to be *brought up* upon the Stage once more ; being only reſerved in *Store* for the Benefit of *Poſterity* :

* *Part* II. *Pag.* 101, 102.

Posterity: and to be, by this *Deterration*, fetch'd out to *Light again* to ſupply the Wants of the *latter Ages* of the *World*. For had theſe *Strata* of Stone, and other mineral Matter, which lay then *underneath*, been altogether deſtitute of this *Vegetative Intermixture*, and had not contained ſome, though a ſmaller and more parſimonious *Supply* of it in them: had there not been alſo vaſt Numbers of *Shells*, *Teeth*, *Bones*, and the like, lodg'd in them, which are, when rotted and diſſolved*, a proper and natural *Manure* to the *Earth*, as all Parts whatſoever of Animals, as well as Vegetables, are; they conſiſting of ſuch Matter as the upper and vegetative *Stratum* it ſelf contains, and therefore ſuch as is fit for the Conſtitution of Plants and of Animals*; I ſay, had it not been for *this*, when the *upper* and vegetative *Stratum* was once *waſh'd off*, and *born down* by Rains, the Hills would have become all perfectly *barren*, the *Strata* below yielding only meer *ſteril* and mineral Matter, ſuch as

* *Vid. Conſect.* 5. *ſupra.*

* *Confer Conſect.* 1. *ſupra.*

was wholey unfit and improper for the Formation of *Vegetables*. Nay, the *Incovenience* would not have ſtop'd *there*, but have ſpread it ſelf much *further*. For, had the Vegetative *Stratum* been carry'd off, the *Devolution* ſtill continued, and ſo the Matter of the *lower* or *mineral Strata* been likewiſe by Degrees *born down* ſucceſſively to the *Roots* and *Bottoms* of the *Hills*, and upon the neighbouring Parts of the *Valleys* and *Plains*, this would, as far as it reach'd, have *cover'd* and *bury'd* the *upper* and *vegetative Stratum* that was expanded over thoſe Valleys and Plains, and render'd as much of them as it ſo cover'd alſo fruſtrate, *ſteril*, and *unfruitful*. So that, by this Means, in the *latter Ages* of the *World*, when the *Earth* ſhould be fully *peopled*, and all Quarters and Corners of it *ſtock'd* with *Inhabitants*, and when conſequently there would be the *greateſt Need* and Occaſion for its *Productions* every where, for Supply of the Neceſſities of theſe its *numerous Inhabitants*, there would have

have been then *much fewer* than ever; a great Part of the Earth being render'd intirely *barren.* So that they might have e'en *ſtarv'd*, had it not been for this Providential *Reſerve:* this *Hoord*, if I may ſo ſay, that was ſtowed in the *Strata* underneath, and now ſeaſonably *diſcloſed* and brought forth.

PART VI.

Concerning the State *of the* Earth, *and the* Productions *of it*, before *the* Deluge.

THE Thread of this *Diſcourſe* draws now near to an *End:* and I have Reaſon to fear that, by this Time, the Reader, as well as my ſelf, thinks it high Time that it were quite ſpun out. For which Reaſon I ſhall not any longer preſume upon his Patience farther than needs I muſt.

In the five foregoing *Parts* of this *Eſſay* I lay down what I have to propoſe relating to the *Condition* of the *Earth* during the *Time* of the *Deluge*, and *ever ſince* that Time. And

And here I am to make a Stand: to look a great way back: and make ſome *Reflections* upon the Poſture of Things *before* the *Deluge.*

The Method I take may perhaps be cenſur'd by ſome as prepoſterous, becauſe I thus treat laſt of the *Antediluvian Earth,* which was firſt in Order of Nature. But they will, I hope, let fall that Cenſure, when they are acquainted that 'tis a thing of Conſtraint, and not of Choice: and that 'twas for want of Footing, and Ground to go upon, that I did not take *that Earth* under Conſideration ſooner. The Truth is, there was no Way for me to come to any competent Knowledge of it, or to give any ſure Judgment concerning it, but meerly by *Induction:* and by Contemplation of the *Shells, Bones,* and other *Remains* of it, which are ſtill in being. Now, before I could inferr any thing from *theſe,* it lay upon me to make out that they all belonged to *that Earth,* and were the genuine *Products* of it: to ſhew likewiſe how they became bury'd and diſpoſed in the Manner we at this Day find them: and by what Means they were preſerved till now. And *that* is what I have been hitherto about; ſo that this is indeed but the Place for this Diſquiſition concerning the *Antediluvian Earth:* and it could not well have been brought in before.

Had there not been *ſtill remaining* a great many *Animal* and *Vegetable Bodyes* that were the legitimate Off-ſprings of *that Earth*, 'twould have been an extravagant and impracticable Undertaking to have gone about to have determined any thing concerning it: and the more ſo because the *Earth* it ſelf was *diſſolved* and *deſtroy'd**. But I prove that there are ſuch *Remains* of it, inclos'd in great Plenty in the Marble, Stone, and the other compacter *Strata* of the *preſent Earth*, whereby they have been preſerv'd, thro' ſo many Ages, quite down to our Times: and are like to endure, many of them, much longer; even as long as the *Strata* themſelves continue in the *Condition* they now are. So that *theſe* will be a ſure and laſting *Monument*, and *Evidence*, to *Poſterity*, quite down to the *End* of

* *Part* II. *Conſect.* 2.

of the *World*, of the *Truth* and *Certainty* of that extraordinary Accident, the *Destruction* of the *Earth*, and of *Mankind*, by the *Deluge*.

Now becaufe the *Obfervations* which I make ufe of in the former Parts of this Work give an Account of the faid *Productions* thus preferved, I proceed upon thofe Obfervations, as hitherto: and, by *Inferences* which eafyly, clearly, and naturaly flow from them, fhew what was the *Condition* and State of *that Earth*, and wherein it differ'd from *this* we now inhabit.

And in regard that, from a *Theory* which, how much foever it may relifh of Wit and Invention, hath no real *Foundation* either in *Nature* or *Hiftory*, the *Author* fo often mention'd already hath fet forth an *imaginary* and *fictitious Earth*, whofe *Pofture* to the *Sun* he fuppofes to have been much *different* from *that* which the *Earth* at *prefent* obtains, and *fuch* that there could be no *Alteration* of Heat and Cold, no Summer and Winter, as now there is, but a conftant *Uniformity* of *Weather*, and *Equality* of *Seafons* *: an *Earth* without

* *Theory of the Earth*, l. 1. c. 6. & l. 2. c. 3.

out any *Sea*: without *Mountains*, or other *Inequalities* *: and without either *Metalls* or *Minerals* †: in few Words, *one* perfectly unlike what the *Antediluvian Earth* was in Truth and Reality: and perfectly unlike that which *Moses* hath represented; I shall interpose some *Consectaries* which would have been otherwise needless and superfluous: which are directly levell'd against these *Mistakes*: and evince that, wherever he hath receded from the *Mosaick Account* of *that Earth*, he hath at the same time also receded from *Nature*, and Matter of *Fact*; and this purely from the aforesaid *Observations*; from which I shall prove,

* Ibid. l. 1. c. 5.
† l. 2. c. 6.

1. That the *Face* of the *Earth*, before the Deluge, was *not smooth*, *eaven*, and *uniform*: but *unequal*, and distinguish'd with *Mountains*, *Valleys*, and *Plains*: as also with *Sea*, *Lakes*, and *Rivers*.

2. That the *Quantity* of *Water* upon the Surface of the Globe was nearly the *same* as now: the *Ocean* of the same *Extent*, and possess'd an equal Share of the *Globe*; intermixing

mixing with the *Land* ſo as to chequer it into *Earth* and *Water*, and to make much the ſame *Diverſities* of *Sea* and *Land* as we behold *at preſent.*

That the *Water* of the *Sea* was ſaturated with *Salt*, in like Manner as *now* it is. That it was agitated with *Tides*, or a *Flux* and *Reflux*: with *Storms* and other *Commotions*. 3.

That the *Sea* was very abundantly *repleniſh'd* with *Fiſh* of all *Sorts*: as well of the cartilaginous and ſquamınoſe, as of the teſtaceous and cruſtaceous Kinds: and that the *Lakes* and *Rivers* were as *plentifully* furniſh'd with Lake and River-Fiſh of *all Sorts.* 4.

That the *Earth* was very *exuberantly* beſet with *Trees*, *Shrubs*, and *Herbs*: and ſtock'd with *Animals*, of all Kinds, *Quadrupeds*, *Fowls*, and *Inſects*: and this on all *Sides*, and in all *Parts* of it, quite round the *Globe.* 5.

That the *Animal* and *Vegetable Productions* of the *Antediluvian Earth* did not in any wiſe *differ* from thoſe of the *preſent Earth.* That there were *then* the very *ſame Kinds* of Animals 6.

Animals and Vegetables, and the ſame ſubordinate *Species* under each *Kind* that *now* there is. That they were of the *ſame Stature* and *Size*, as well as of the *ſame Shape*: their *Parts* of the *ſame Fabrick*, *Texture*, *Conſtitution*, and *Colour*, as are thoſe of the Animals and Vegetables at *this Day* in being.

7. That there were *Metalls*, and *Minerals*, in the *Antediluvian Earth*.

8. That the Terraqueous *Globe* had the *ſame Site* and *Poſition* in reſpect of the *Sun* that it *now* hath. That its *Axis* was not *parallel* to that of the *Ecliptic*, but *inclined* in like Manner as it is at *preſent*: and that there were the ſame *Succeſſions* of *Heat* and *Cold*, *Wet* and *Dry*: the ſame *Viciſſitudes* of *Seaſons*, Spring, Summer, Autumn, and Winter, that *now* there is.

It hath been already noted, that theſe *Propoſitions* are founded on *Obſervations* made on the Animal and Vegetable *Remains* of the *Antediluvian Earth*. From thoſe Remains we may judge what Sort of Earth *that* was: and ſee that it was not different from *this* we now inhabit.

Now

Now though 'tis not to be expected that I *here* formally lay down those *Observations*, that being not the Business of *this Tract*, yet until I have Opportunity both of doing so, and of shewing in what Manner the foregoing *Propositions* flow from them, it may be very convenient that I give some short *Directions* how the Reader, for his present Satisfaction, may, of himself, and without my Assistance, make out the principal Articles of these *Propositions* from the *Observations* already deliver'd in the several Parts of this Discourse, and from one or two more that I shall add upon this Occasion. And that he may at one View discover how consonant the *Account* which *Moses* hath left us of the Primitive Earth, is to *this* which we have from *Nature*, and how much the late *Theory of the Earth* differs from both, I will set down that Writer's Sense of the Matter under each Head as we pass along.

To begin therefore with the *Sea*; That there was *one* before the Deluge, there needs not, I think, any other Proof than the *Productions* of it it yet in being: the *Shells*, the *Teeth*, and *Bones* of *Sea Fishes* *. And for *Moses*, he is not at all averse hereto; but as expresly asserts that there was then a *Sea*, as the *Theory* does that there was none. Take it in his own Words. **And God said, Let the Waters under the Heaven be gathered together unto one Place, and let the Dry-land appear: and it was so. And God called the Dry-land Earth, and the gathering together of the Waters called he* SEAS: *and God saw that it was good.* Though the *Theorist* flatly denyes that there was then any such Thing, yet he does not go about to dispute the Translation of this Passage, but readily owns*, that *Moses* hath here *used a Word that was common and known to signify the Sea.* According to *him* therefore, we see the *Sea* was formed at the *Beginning* of the *World*, and after its Formation approved of as *good:* that is, very necessary and serviceable to the Ends of Providence in the Kingdom of Nature. Which indeed it is so many Ways, that it must needs be granted that *that* would have been

* *Confer* Part II.

* Gen. 1. 9, 10.

* L. 1. c. 7.

been a very wild World had it been
without any Sea. The *separating*
of the *Sea* and *Land*, and determining the set *Bounds* of each, is
here * reckon'd Part of the Work * Vers. 13.
of the third Day: as the stocking
of the *Sea* with *Whales* and other
Fishes, is * of the fifth. *And God* * Vers. 21,
created great Whales, &c. *and bles-* 22.
sed them, saying, be fruitfull and
multiply, and fill the Waters in the
SEAS. And when, on the sixth
Day, the finishing Hand was set to
the Work, and Man created, God
gives him *Dominion over the Fish of*
the SEA *. 'Twould have been but * Vers. 28.
a scanty and narrow Dominion, and
Adam a very mean Prince, had there
then been neither any *Fish* existent,
nor *Sea* to contain them. Nay,
this had been little better than a
downright Illusion and abusing of
him: and what is more, *that World*
had been so far from excelling *ours*
in the Abundance of its *Productions*,
which is what the *Theorist* contends
for on another Occasion, that 'twould
have fallen far short of it: have
wanted a very noble and large
Share of the Creation which we
enjoy:

enjoy: been deprived of a moſt excellent and wholeſome Fare, and very many delicious Diſhes that we have the Uſe and Benefit of. But the Caſe was realy much otherwiſe: and we have as good Proof as could be wiſh'd that there were not any of theſe wanting. The Things, many of them yet extant, ſpeak aloud for themſelves: and are back'd with an early and general *Tradition*. For *Moſes* is ſo far from being ſingular in thus relating that the Sea is of as *old* a *Date* and *Standing* as the *Earth* it ſelf, that he hath all, even the firſt and remoteſt *Antiquity* of his Side; the *Gentil Accounts* of the Creation making the *Ocean* to riſe out of the *Chaos* almoſt as ſoon as any Thing beſides. But we have in Store a yet further *Teſtimony* that will be granted to be beyond all Exception. 'Tis from the Mouth of *God* himſelf, being Part of the *Law* promulgated by him in a moſt ſolemn and extraordinary Manner. Exod. 20. 11. *In ſix Days the Lord made Heaven and Earth, the SEA, and all that in them is.* 'Tis very hard to think the *Theoriſt* ſhould not

not know this: and as hard that, knowing it, he ſhould ſo openly diſſent from it.

Then for the *Dimenſions* of the *Sea*; that it was as *large*, and of as great *Extent* as now it is, may be inferr'd from the vaſt *Multitudes* of thoſe *marine Bodyes* which are ſtill found in *all Parts* of the known World*. Had *theſe* been found in only one or two Places: or did we meet with but ſome few Species of them, and ſuch as are the Products of one Climate or Country, it might have been ſuſpected that the Sea was then, what the *Caſpian* is, only a great Pond or Lake, and confined to one Part of the Globe. But ſeeing they are dig'd up at *Land* almoſt *every where*, and in at leaſt as great Variety and Plenty as they are obſerv'd at *Sea*: ſince likewiſe the *foſſil Shells* are many of them of the *ſame Kinds* with thoſe that *now* appear, if not on the neighbouring, upon remoter *Shores*, and *the reſt* ſuch as may well be preſumed to be living at the Bottom, or in the interiour and deeper Parts of the *adjacent Seas* *, we may reaſonably

* *Confer pag.* 6. & *Part* II.

* *Confer Pag.* 25 *ſupra.*

nably conclude, not only that the *Sea* was of the *ſame Bigneſs* and *Capacity* before the *Deluge*, but that it was of much the *ſame Form* alſo, and interwoven with the *Earth*, in like Manner as at this Time : that there was *Sea* in or near the very *ſame Places* or Parts of the Globe : that *each Sea* had its *peculiar Shell-Fiſh*, and thoſe of the *ſame Kinds* that now it hath : that there was the *ſame Diverſity* of *Climates*, here *warmer* and more agreeable to the *Southern Shell-Fiſh*, there *colder* and better ſuited to the *Northern* ones : the ſame *Variation* of *Soils*, *this Tract* affording ſuch a Terreſtrial Matter as is proper for the Formation and Nouriſhment of *one Sort* of Shell-fiſh, *that* of *another* : in few Words, that there was then much the ſame *Appearance* of *Nature*, and *Face* of *Things* that we behold in the *preſent Earth*. But of this more by and by.

That the *Water* of the *Sea* was *ſalt*, as *now* it is, may be made out likewiſe from thoſe *Shells*, and other the *Productions* of it. For they are of the ſame *Conſtitution*, and conſiſt of the ſame Sort of *Matter* that do the

the Shells at *this Day* found upon our *Shores* *. And particularly ſuch of them as remain ſound and unperiſh'd yield, upon Tryal, a true *Marine Salt*; in like Manner as *theſe* alſo do. The *Salt*, wherewith the *Sea-Water* is ſaturated, is part of the *Food* of the *Shell-Fiſh* reſiding therein, and a main *Ingredient* in the Make of their *Bodyes*; they living upon *this*, and upon the *Mud* and other vegetative *Earthy Matter* there. Now that *that Sea* was *Salt*, there needs not I think a fuller Proof than that the *Shells* which are owing to it thus retain ſtill in them a real *marine Salt*.

* *Vid. pag.* 23, 24, & 25, *ſupra.*

And that the Sea *ebb'd and flow'd* before the Deluge, may be inferr'd, not ſo much from the Neceſſity of *that Motion*, and the many and great *Uſes* of it in the Natural World *, as from certain *Effects* that it had upon the *Shells*, and other like *Bodyes* yet preſerved. 'Tis known that the Sea, by this Acceſs and Receſs, ſhuffling the empty *Shells*, or whatever elſe lies expos'd upon the Shores, and bearing them along with it backwards and forward upon

* *Confer Pag.* 51, & 173.

on the Sand there, *frets* and *wears* them away by little and little, in tract of Time reducing those that are concave and gibbose to a flat, and at length *grinding* them away almost to nothing. And there are, not uncommonly, found *Shells* so *worn* enclos'd, amongst others, in *Stone*.

As the Sea-Shells afford us a sure Argument of a Sea, so do the *River-Shells* of *Rivers*, in the *Antediluvian* Earth. And if there were *Rivers*, there must needs also have been *Mountains*; for *they* will not *flow* unless upon a *Declivity*, and their Sources be rais'd above the Earth's ordinary Surface, so that they may run upon a *Descent**; the *Swift-*

* *Confer Part* 3. *Sect.* 1. *Pag.* 170, & 171, *supra.* The *Theorist*, I know, supposes both the Antediluvian and the present Earth to be of an Oval Figure, and protended towards the Poles; as thinking that such a Figure would afford him a Plane so much inclined towards the Æquator, that the Rivers might flow upon it though there were no Mountains. But 'tis certain they could not. Nor are there any the least Grounds to believe that the first Earth was of that Figure. If he had had any thing that had look'd like a Proof of it, he had done well to have produced it. But 'tis evident, though we imagine the Earth formed that Way he proposes, it would not have fallen into any such Figure. And for the present Earth, 'tis of a Figure as different from that which he assigns as it well could be; it being a *Spheroides prolatus*, as appears from the late Discoveries concerning it.

Swiftneſs of their *Current*, and the *Quantity* of *Water* refunded by them, being proportioned generaly to the *Height* of their *Sources*, and the *Bigneſs* of the *Mountains*, out of which they ariſe. *Mountains* being proved, nothing need be ſaid concerning *Valleys*; they neceſſarily following from that Proof, as being nothing but the Intervalls betwixt the Mountains. But let us ſee what *Moſes* hath on this Subject *, *And the Waters* (he is treating of the Deluge) *prevailed exceedingly upon the Earth: and all the* HIGH HILLS *that were under the whole Heaven were cover'd. Fifteen Cubits upwards did the Waters prevail: and the* MOUNTAINS *were cover'd. And all Fleſh dyed: — all in whoſe Noſtrils was the Breath of Life.* The *Theoriſt* averrs that there were *no Mountains* in the *firſt Earth*: and therefore would have *this* to be underſtood of *thoſe* which were raiſed *afterwards*. But that cannot be. For the Hiſtorian here plainly makes *theſe Mountains* the *Standards* and *Meaſures* of the *Riſe* of the *Water*; which they could never have been

* *Gen.* vii. 19. & *ſeqq.*

had they not been *ſtanding* when it did ſo *riſe* and overpour the *Earth*. His Intention in the whole is to acquaint us that all Land-Creatures whatever, both Men, Quadrupeds, Birds, and Inſects, periſh'd, and were deſtroy'd by the Water, *Noah* only excepted, *and they that were with him in the Ark.* And at the ſame Time, to let us ſee the *Truth* and *Probability* of the Thing: to convince us that there was no Way for any one to eſcape, and particularly that none could ſave themſelves by climbing up to the Tops of the *Mountains* that then were, he aſſures us that *they*, even the higheſt of them, were all *cover'd* and bury'd under *Water*. Now to ſay that there were then *no Mountains*: and that *this* is meant of *Mountains* that were not formed till *afterwards*, makes it not intelligible, and indeed hardly common Senſe.

The extreme *Fertility* of both Sea and Land before the Deluge, appears ſufficiently from the vaſt and almoſt incredible *Numbers* of their *Productions* yet extant*; not to inſiſt upon thoſe which are long ago rotted

* *Vid.* *Part* II.

rotted and gone †. Nor need we much wonder at this their *abundant Fruitfulneſs*, when we know from what Sourſe it proceeded; which our Hiſtorian hath opened to us in very ſignificant Words ‖. *And God ſaid, let the Waters bring forth abundantly the moving Creature that hath Life*, &c. — *And God bleſſed them, ſaying, be fruitfull and multiply, and fill the Waters in the Seas: and let Fowl multiply in the Earth, &c.* Here was, we ſee, a Bleſſing, handed out with the firſt Pairs of Animals at the Moment of their Creation, very liberal and extenſive: and it had Effect with a Witneſs. A Man that does but behold the *mighty Sholes* of *Shells*, to take *them* for an *Inſtance*, that are ſtill remaining, and that lye bedded and cumulated in many Places *Heap* upon *Heap*, amongſt the Ordinary Matter of the Earth, will ſcarcely be able to believe his Eyes, or conceive which Way *theſe* could ever live or ſubſiſt one by another. But yet ſubſiſt they did: and as they themſelves teſtify, well too; an Argument that *that Earth* did not deal out their *Nouriſhment*

† *Confer Pag.* 34, 70, 71, 89, & *Pt.* III. *Sect.* 2. *Conſ.* 11.

‖ *Gen.* i. *v.* 20. & *ſeq.*

with an over-ſparing or illiberal Hand.

That theſe *Productions* of the *Original Earth*, *differ* not from thoſe of the *Preſent*, either in *Figure*, in *Magnitude*, in *Texture*, or any other Reſpect, is eaſyly learn'd by comparing of them. The exact *Agreement* betwixt the *Marine Bodyes* I have ſhewn already*: and ſhall in due Place ſhew the ſame of the *terreſtrial ones*.

* *p.* 23, 24, 25. *Conf. Gen.* i. 11, 12, 21, 24, 25. & vi. 20.

As there were ſuch great Numbers of Animals and Vegetables in the *Primitive* Earth, ſo that there were alſo *Metalls* and *Minerals*, and theſe in no leſs *Plenty* than in *ours*, is very clear from what hath been deliver'd in the *Fourth Part* of *this Eſſay*, which need not be repeated here. Nor is *Moſes* defective in this Point*. *And Zillah, ſhe alſo bare Tubal-Cain, an Inſtructer of every Artificer in Braſs and Iron.* The *Theoriſt*, quite contrary, ſays, *As for Subterraneous Things, Metalls and Minerals, I believe they had none in the firſt Earth; and the happier they; no Gold, nor Silver, nor coarſer Metalls.* Amongſt theſe *coarſer Metalls*

* *Gen.* iv. 22.

talls are *Copper*, or *Braſs*, and *Iron*. Now if there were *none* of theſe, 'tis a great Myſtery to me, I confeſs, how *Tubal-Cain*, who certainly dy'd either before or at the Deluge*, could ever have taught the Workmanſhip and Uſe of them. And yet if this *Theory* be true, there neither was nor could be any within their reach, or that they could ever poſſibly come at. For the Truth of the *Theory* I am in no wiſe concerned; the *Composer* of it muſt look to that; but that there were realy both *Metalls* and *Minerals* before the *Deluge*, is moſt certain. For beſides the Teſtimony that we have of the Thing from *Nature*, and the Paſſage already alledged out of *Moſes*, there is *another* for which we are alſo obliged to the ſame Author, that acquaints us there were *both* even in *Paradiſe* it ſelf. 'Tis in his ſecond Chapter*. *The Name of the firſt* River *is Piſon: that is it which compaſſeth the whole Land of Havilah, where there is Gold. And the Gold of that Land is good: there is Bdellium and the Onyx-ſtone.* He ſpeaks here, I grant, only in the Preſent

* *Confer Gen.* vii. 23. & 1 *Pet.* iii. 20.

* *Gen.* ii. 11, 12.

Preſent Tenſe, *there is Gold:* but muſt mean not only that there was Gold and Gemms there in *his Time*, but that there was ſo likewiſe from the *Beginning* of the *World*, of which he is giving an Account in theſe two Chapters, or, with Submiſſion, I conceive 'twould not be any thing to his Purpoſe. He is here ſpeaking of *Paradiſe*; which he repreſents as a moſt charming and delightful Place, beſet with *every Tree that is pleaſant to the Sight, and good*
† Verſe 9. *for Food*†*:* water'd with refreſhing Streams, and excellent Rivers: and abounding with Things, not only uſeful and convenient, but even the moſt rare and valuable, the moſt coſtly and deſirable; particularly *Gold*, *Precious Stones*, and *Perfumes*; which were all much eſteem'd and admir'd by the *Jews*, to whom he wrote this. Nor is it any Paradox, notwithſtanding that *Diſſolution* of the *Earth* which happen'd at the
Part II. *Deluge**, to ſuppoſe there was this or that *Metall* or *Mineral* in the *ſame Part* of the *Globe* afterwards where it was before *that* happen'd. The *Water* of the *Abyſs* indeed changed its

its *Place*, during the Time. So did the *Sea*, and bore the *Bodyes* it contained, many of them out along with it. But for the *terreſtrial Parts* of the Globe, *Metalls*, *Minerals*, Marble, Stone, and the reſt, they, tho' diſſolv'd, and aſſumed up into the Water, did not *flit* or *move* far; but, at the general Subſidence, ſettled down again in or near the *ſame Place* from which they were before taken up. For the *Water* was all out upon the Face of the Earth before ever *theſe* ſtirr'd, or were fetch'd up out of their native Beds: and they were all ſunk down into the ſame Beds again, before the Water began to ſhift away back to its old Quarters; ſo that it could not contribute any thing to the Removal of them. Even the very *Vegetables*, and their *Seeds*, which were many of them naturaly lighter than the Water, aſſiſted by the heavier terreſtrial Matter that had in this Jumble and Confuſion faſten'd and ſtuck to them, fell all to the Bottom: and the *Water* was in great meaſure *clear*, and diſengag'd from the *Earthy Maſs*, before it went off.

And

And 'twas well it was ſo; for had the *Mineral Matter* of the *Globe* not been held to its *former Station*, but hurried about and *tranſpos'd* from Place to Place, 'tis ſcarcely to be conceiv'd how many and great *Inconveniences* it would have occaſion'd. The ſame likewiſe for *Vegetables*. Had the *Seeds* of the *Pepper Plant*, the *Nutmeg*, the *Clove*, or the *Cinnamon Trees*, been born from *Java*, *Banda*, the *Moluccoes*, and *Ceylon*, to theſe *Northern Countryes*, they muſt all have *ſtarved* for want of *Sun*. Or had the *Seeds* of our *colder Plants* ſhifted thither, they would have been burnt up and ſpoil'd by it. But *Things* generaly kept to their proper Places: to their old natural Soil and *Climate*; which had they not done, all would have been confounded and deſtroy'd. 'Tis true, the *Vegetables*, being comparatively *lighter* than the ordinary terreſtrial Matter of the Globe, ſubſided *laſt**: and conſequently lying many of them upon the *Surface* of the *Earth*, thoſe which were of conſiderable *Bulk*, as the bigger Sorts of *Trees*, which had large and ſpreading

* *Confer Part* II. *Conſect*. 3.

ſpreading Heads, would lye with their Branches ſtretch'd up to a great Height in the *Water*, (and, when *that* was withdrawn, in the Air *,) and ſo, being very much in the *Water*'s. Way, when it began to depart and retire back again, would be apt to be *remov'd* and driven forward along with it, eſpecialy thoſe which lay in ſuch Places where the *Current* happened to run ſtrong. Accordingly we *now* find of *theſe Trees* in the *Northern Iſlands*, and the other bleaker and *colder Parts* of the *Earth*, where none *now* do, or perhaps ever did, *grow* *. And *there* they are of mighty *Uſe* to the *Inhabitants*, affording them a Supply of *Timber* which their own Country doth not yield, and which they employ not only for *Fewel*, which yet is much needed in thoſe *cold Countryes*, but for *Building* likewiſe, and many other *Purpoſes*. Whereas in the *Places* whence they were thus driven, they would have been *uſeleſs*, yea but an *Incumberance*, and might be eaſyly ſpared. For as long as the *Seeds* remained behind, lodged in a natural and agreeable Soil,

* *In which Poſture 'tis probable the* Olive-Tree *lay from which the* Dove *pluck'd off the* Leaf *that ſhe brought unto* Noah, *Gen.* viii. 11.

* *Confer Pag.* 127. *ſupra.*

Soil, all was ſafe enough; *they* would ſoon vegetate, and ſend forth a *new* Sett of *Trees* there, ſo that 'twas not much matter what became of the *Old ones*. But to the *Parts* whereunto the Trees were thus removed, they are of great *Advantage*. And, which is in Truth very remarkable, and an Argument that there was ſomething more than meer *Chance* in this *Affair*, there are hardly any *Countryes*, eſpecialy in the *Northern* and *colder Climes*, that are deſtitute of *Timber* of their *own Growth*, which have not a very large *Supply* of theſe *Stray-Trees*, if I may ſo call them. But to proceed. After that the Terreſtrial Matter was once ſunk down into its ſeveral Beds, and well ſettled there, the *Mountains* were caſt up, and the *Springs* and *Rivers* burſt forth, in ſuch Numbers, and at ſuch Diſtances from one another, in *all Parts* of the Globe, as beſt anſwer'd the Neceſſities and Expences of *each*: and therefore undoubtedly in much the *ſame Places* that they were before the *Deluge*. All Things were ſo contriv'd and order'd in the *Re-fitting*

ting up the *Globe* at this Time, that they might beſt execute and perform each their ſeveral *Ends* and *Offices*. There were the ſame *Meaſures* taken, and the ſame *Proceſs* us'd in this *Re-Formation* of it, that were us'd when 'twas *firſt built* : and much ſuch an *Earth* aroſe out of the *Deluge*, as at the *Creation*, ſprung out of *Nothing**. But the Reader will more clearly diſcover all this, with the Reaſons of it, if he give but himſelf the Trouble to compare *Part* 2. *Conſ*. 2. & *ſeq*. *Part* 3. *Sect*. 1. *pag*. 171. & *ſeq*. & *Sect*. 2. *Conſ*. 2, 3, & 7. *Part* 4. *Conſ*. 3. & *Part* 6. *Conſ*. 9. This premis'd, it would be, I think, not ſtrange ſhould we find *Paradiſe* at this Day where *Adam* left it : the ſame *Rivers* : the ſame *Face* of the *Ground* : the ſame *Metalls* and *Minerals*, that *then* there were. And I the rather note this, becauſe I ſee there are ſome ſo earneſt in Queſt of it. Learned Men have been now a long Time ſearching after the happy Country from which our firſt Parents were exil'd. If they can find it, with all my Heart : and there have been ſome that

* *Vid. Pag.* 109. *ſupra.*

that have ſought it with that Induſtry that I think they deſerve it for their Pains whenever they make the Diſcovery. To deal freely, I am of Opinion there's no Part of the *preſent Earth* that does come up fully to the *Moſaick* Deſcription of *Paradiſe*. The Country about *Babylon*, or *Bagdet*, bids faireſt for it : and I am perſwaded that it was thereabouts. But if ſo, whoever ſhall compare this Country, as *now* it ſtands, with *that* Deſcription, will find that it ſuſtained ſome *Alterations* from the *Deluge*, perhaps *more* than any Part of the *Earth* beſides. And there's an obvious *Reaſon* why it ſhould. There was a Paradiſe *before*, but was to be none *after* the Deluge. The *Caſe* was *alter'd*, and the *Reaſon* of the Thing *ceaſed*. So that *all* that denominated it *Paradiſe*, and that *diſtinguiſh'd* it from the *reſt* of the *Globe*, was *lop'd off* by the *Deluge :* and *that* only left which it enjoy'd in common with its neighbour Countryes. Upon the whole, 'tis, I think, apparent that what I offer in this Diſcourſe is ſo far from doing any Diſkindneſs

to

to the Cauſe theſe Gentlemen are, and have been ſo long, engag'd in, that it does them a real Service, and helps them out with the main Thing whereat they ſtuck; fairly ſolving all *Difficulties* in the *Moſaic* Relation of *Paradiſe.* Wherefore now to proceed to the laſt Head to be diſcuſs'd, the *Viciſſitude* of *Seaſons*, Summer and Winter, Heat and Cold, in the *Antediluvian World.*

And that there realy was ſuch a *Viciſſitude* we need not go any further for Proof than to the aforeſaid *Animal* and *Vegetable Bodyes* ſtill preſerved; the general Tenour of them ſpeaking it out ſo plainly as to leave no Room for Doubt.

There are, we know, ſome Sorts of *Vegetables* which conſiſt of *Particles* very *fine, light*, and *active:* and which therefore require only a *ſmaller Degree* of *Heat* to raiſe them*, from out the Earth, up into the Seeds, Roots, or Bodyes of thoſe Vegetables, for their Growth and Nouriſhment. So that for the Raiſing of *theſe*, the *Sun's Power*, when only *leſſer*, is *ſufficient.* And therefore they begin to appear in the

* *Vid.* *Part* III. *Sect.* 1. *Conſect.* 8. *Pag.* 139.

earlyer Months, in *February*, and *March*; when they firſt peep forth of the Ground: after a while they diſplay themſelves, ſhewing their whole Tire of *Leaves*: then their *Flowers*: next their *Seeds*: and laſtly when, in the following Months, *April*, and *May*, the *Sun* is farther advanc'd, and (to ſpeak in the Phraſe of the *Vulgar*, which I chooſe all along for the Sake of Plainneſs) hath gain'd a greater Strength, the *Heat* becomes too powerful and boiſterous for them; it now mounting up the Terreſtrial Matter with ſuch *Force* and *Rapidity*, that the *Plants* cannot aſſume that Part of it which is proper for their Nouriſhment, as it paſſes through them, nor incorporate it with them, as before they were wont when it paſs'd more gently and leiſurely. Nay the *Heat* at length grows ſo *great*, that it again *diſſipates* and bears off thoſe very *Corpuſcles* which before it *brought*; the *Parts* of theſe *Plants* being very *tender*, as conſiſting of *Corpuſcles* which are extremely *ſmall* and *light*, and therefore the more eaſyly *diſſipable*. So that *then* theſe *Plants dye* away,

away, ſhrink down again into the Earth, and all, ſave only their Roots and Seeds, *vaniſh* and *diſappear.* But when the *Sun's Heat* is thus far *advanc'd*, 'tis but juſt come up to the Pitch of *another Sett* of *Vegetables:* and but great enough to excite and bear up the *Terreſtrial Particles*, which are more *craſs* and *ponderous.* And therefore thoſe *Plants*, which conſiſt of *ſuch*, begin *then* to ſhoot forth, and diſplay themſelves. So that the Months of *April* and *May* preſent us with another *Crop* and *Rank* of *Plants.* For the ſame Reaſon alſo, *June, July*, and *Auguſt* go *farther*, and exhibit a ſtill *different Shew* of *Vegetables*, and Face of Things. But when, in the Months of *September* and *October*, the *Sun's Power* is again *diminiſh'd*, and its *Heat* but about equivalent to *that* of *March* and *April*, it again ſuits the *Plants* which were *then* in Seaſon; ſo that they, many of them, ſpring up afreſh in *theſe* Months, and flouriſh over anew, in like Manner as before they did in *thoſe*; till being check'd by the *Cold* of the ſucceeding *Winter*, the

Sun being gone off, and having now no longer *Power* great enough to bring up and ſupply them with freſh Matter, they preſently begin to *decline* for want thereof, and at length quite dwindle away and *diſappear*, untill the Arrival of the *Spring Seaſon*, when they take their *Turns* over again as formerly. Yea the more *tender* and *fugitive Parts*, as the *Leaves* of many of the more *ſturdy* and *vigorous Vegetables*, *Shrubs*, and *Trees*, ſuffer the ſame Fate, and *fall off* for want of the Supply from beneath; thoſe only which are more tenacious, conſiſtent, and hardy, enduring the Brunt, and making a ſhift to ſubſiſt for the Time without ſuch Recruit and Reparation. 'Tis therefore, we ſee, moſt apparent that this *Succeſſion* of *Things* upon the Face of the Earth, is purely the Reſult and Effect of the *Viciſſitude* of *Seaſons*: and is as conſtant and certain as is the *Cauſe* of that *Viciſſitude*, the *Sun's Declination*: ſo certain, that were a Man kept for ſome Time Blindfold, in ſuch Manner that he could have no Notice how the *Year* paſs'd paſs'd on, and was at length turn'd forth into the next Field or Garden, he would not need any other Almanack to inform him what *Seaſon* of the Year it then was.

But if, inſtead of this *Variation* of *Heat*, we ſuppoſe that there was an *Equality* or conſtant *Temperature* of it before the *Deluge*, which is what the *Theoriſt* contends for, the Caſe would be very much alter'd, and *that* altogether for the worſe. A Man can hardly at firſt imagine what a *Train* of ill *Conſequences* would follow from ſuch a Condition and Poſture of Things; of which 'twould not be the leaſt that ſuch a Mediocrity of Heat would deprive the World of the moſt beautiful and the moſt uſeful Parts of all the whole Creation: and would be ſo far from exalting the Earth to a more happy and Paradiſiacal *State*, which is what he brought it in for, that 'twould turn it to a general *Deſolation*, and a meer barren Wilderneſs, to ſay no worſe. Such an *Heat* would be too *little* for *ſome Sorts* of *Vegetables*, and too *great* for *others*. The more *fine* and *ten-*

der Plants, thoſe which will not bear a Degree of *Heat* beyond that of *April*, would be all *burnt* up, and *deſtroy'd* by it; whilſt it could never reach the more *lofty* and *robuſt*, nor would there be near *Heat* enough to ripen their *Fruits* and bring them to *Perfection*. Nothing would ſuit and hit *all*, and anſwer every *End* of *Nature*, but ſuch a gradual *Increaſe* and *Decreaſe* of *Heat* as *now* there is. I muſt not deſcend to the *Animal* World, where the *Inconveniencies* would be as many and as great as in the *Vegetable*: and ſuch a Situation of the *Sun* and *Earth*, as that which the *Theoriſt* ſuppoſes, is ſo far from being preferable to *this* which at preſent obtains, that this hath infinitely the *Advantage* of it in all Reſpects.

Be that how it will, for I have no need to inſiſt upon it, but may take the Thing in his Way, and ſuppoſe that ſuch a *Temperature* would have all the happy Effects that he expects from it; yet there is one very conſiderable *Phænomenon* of the Vegetable Remains of *that Earth*, which affords us a ſure and plain Indication

Indication that there was not then any ſuch *Temperature.* From *theſe* it clearly appears that there was the ſame *Order* and *Succeſſion* of Things upon the *Face* of the *Earth* that there is at *this Time.* Now this *Succeſſion* being, as we have ſeen, cauſed meerly by the *Variation* of the *Sun's Heat*, it muſt needs follow, that there was *then* the *ſame Variations* of it, and conſequently the ſame *Alternations* of *Seaſons*, that *now* there is. Had there been an *Equality* of *Heat*, if we grant that it could have produced all the *Plants* in Nature, which 'tis impoſſible it ever ſhould, it muſt have done it *indifferently* and *uncertainly.* There could be no Reaſon why they ſhould flouriſh at any *one ſet Time* rather than *another.* That is peculiarly the Effect of the *Sun's Variation.* So that they muſt needs have been all in *Confuſion :* and this *Succeſſion* of *Things* would have been quite overturned. The *Plants*, which now appear in the moſt *different* and *diſtant Seaſons*, would have been all in *Prime*, and flouriſhing *together* at the *ſame Time.* So that they would have had *February* and *May*, *July* and *September*, all in one Scene. Nay, the ſeveral *Individuals* of the *ſame Kind* muſt have been as greatly at odds: one arrived to *Seed*, and that fully *ripe*, and ready to *ſhed*; whilſt another was not ſo much as come to *Flower*, but in as differing a *State* and *Hue* as could be. In brief, there would have been all the *Diverſity*, *Uncertainty*, and *Diſorder*, in the *Vegetable Kingdom* that can well be conceiv'd. Which indeed is no more than what he freely owns; telling us that then *Every Seaſon was a Seed-time to Nature. and every Seaſon an Harveſt.* This is what he does, and muſt grant: and this is as much as is needful for the Overturning *his Hypotheſis*. For the *Vegetable Remains* of *that Earth* ſay no ſuch Thing: they give not any the leaſt Countenance to theſe *Conjectures*, but the quite contrary: and *theſe*, being many of them encloſ'd in very fine and cloſe Stone, are *preſerved* to this Day very curiouſly, and entire, to Admiration. By *them* we may eaſyly judge how Things *then* ſtood. And

And there is ſo great an Uniformity, and general Conſent amongſt them, that from it I was enabled to diſcover what *Time* of the *Year* it was that the *Deluge* began *; the whole *Tenour* of *theſe Bodyes* thus preſerv'd clearly pointing forth the Month of *May* †. Nor have I ever met with ſo much as one ſingle *Plant*, or other *Body*, amongſt all thoſe vaſt *Multitudes* which I have carefully view'd, that is peculiar to any *other Seaſon* of the *Year*: or any thing that falls out earlier, or later:

† *Confer Part* III. *Sect.* 2. *Conſect.* 5.

---

* *Gen.* vii. 11. *In the ſecond Month, the ſeventeenth Day of the Month,—were all the Fountains of the great Deep broken up, and the Windows of Heaven were opened.* *Moſes*, writing to the *Jews*, his Country-men, makes uſe of the Form of the Year then receiving among them, which was indeed the firſt and moſt ancient, but had been diſuſed during the Time of their Abode in *Egypt*, and but newly re-eſtabliſh'd when this was wrote. [*Exod.* xii. 2.] In *this*, *Niſan*, or, as 'twas alſo call'd, *Adid*, was the firſt Month: and *Ijar* the ſecond; upon the 17th Day whereof the Waters of the Deluge came forth, according to this Relation. And truly that Time (which is not a little remarkable). falls within the Compaſs here chalk'd out by Nature ſo very punctualy, that one can ſcarcely forbear concluding that theſe Strokes and Lines of *Nature*, and thoſe of *that Relation*, came both from the *ſame Hand*; but this only by the By. The Particulars of the Computation I here uſe ſhall be given at Full elſewhere, they being too many for this Place.

later: any of them ſhort, or any further advanc'd in Growth, Seed, or the like, than they now uſualy are in *that Month*; which aſſuredly could never have happen'd, had there realy been ſuch an *Equality* of *Seaſons*, and conſtant *Temperature* of Heat, as is imagined by the *Theoriſt*. There are ſome *Phænomena* of the *Animal Remains* of *that Earth* which afford us more Arguments to the ſame Purpoſe, and thoſe not leſs concluding than the others; but theſe I ſhall wholey wave for the preſent, there being indeed no Occaſion to make uſe of them here.

I ſhall now only look a little into the *Moſaic* Archives, to obſerve what they furniſh us with upon this Subject, and I have done; for I perceive I have, before I am aware, much exceeded the Meaſures I deſign'd; which, on ſo copious a Subject, 'twas hard not to do. *Gen.* i. 14. *And God ſaid, let there be* LIGHTS *in the Firmament of the Heaven, to divide the Day from the Night: and let them be for Signs, and for* SEASONS, *and for Days, and Years.* This Paſſage, we ſee does not at all favour the

the Opinion that there was no *Variation* of *Seaſons* before the *Deluge.* So far from it, that ſhould a Man go about with never ſo ſet Study and Deſign to deſcribe ſuch a Natural *Form* of the *Year* as is that which is at preſent eſtabliſh'd, he could ſcarcely ever do it in ſo few Words again that were ſo fit and proper, ſo full and expreſs; eſpecialy if, by *Signs,* in this Place, *Months* are intended; for then we have here, firſt the *Year:* and that ſubdivided into its uſual Parts, the four Quarters or *Seaſons,* the twelve *Signs* or Months, and *Days.* Nay at the ſame Time, from the 19th Verſe, we learn that *this Eſtabliſhment* is, within four Days, as old as the World. But further, *Gen.*viii. 21, 22. *And the Lord ſaid in his Heart, I will not again curſe the Ground,— neither will I again ſmite any more every Thing living as I have done. While the Earth remaineth, Seed-Time and Harveſt, and Cold and Heat, and Summer and Winter, and Day and Night, ſhall not ceaſe.* This was pronounc'd upon *Noah*'s Sacrificing, at his coming forth of the Ark,

Ark, after the *Deluge* was over: and implies, that there had indeed then lately been a mighty *Confusion* of Things, for the Time: an Interruption and Perturbation of the *ordinary Course* of them: and a Cessation and Suspension of the *Laws* of *Nature:* but withall gives Security and Assurance that there should never be the like any more to the End of the World: that for the future they should all run again in their old Chanel: and that particularly there should be the same *Vicissitudes of Seasons, and Alternations of Heat and Cold,* that were *before the Deluge.*

FINIS.

BOOKS wrote by the AUTHOR

BRIEF Inſtructions for making Obſervations in all Parts of the World: as alſo for collecting, preſerving, and ſending over Natural Things. Being an Attempt to ſettle an univerſal Correſpondence, for the Advancement of uſefull Knowledge, both Natural, and Civil, 4*to.*

Naturalis Hiſtoria Telluris illuſtrata & aucta. Unà cum ejuſdem Defenſione; præſertim contra nuperas Objectiones D. El. Camerarii Med. Prof. Tubingenſis. Accedit Methodica, & ad ipſam Naturæ Normam inſtituta, Foſſilium in Claſſes Diſtributio. 8*vo.* 1714.

The State of Phyſick, and of Diſeaſes; with an Idea of the Nature and Mechaniſm of Man, of the Diſorders to which it is obnoxious, and of the Method of Rectifying them, 8*vo.* 1718. Theſe three printed for *R. Wilkin*, at the *King's Head* in *St. Paul's Church-Yard.*

Some Thoughts and Experiments concerning Vegetation. Read in a Lecture before the Royal Society: and afterwards printed by Command of the ſaid Society. *Philoſ. Tranſ.* Vol. 21. N° 253.

Remarks upon the antient and preſent State of *London*, occaſion'd by ſome *Roman* Urns, Coins, and other Antiquities, lately diſcover'd. The Third Edition. Printed for *A. Betteſworth* and *W. Taylor* in *Pater-noſter Row*, *R. Goſling* in *Fleet-ſtreet*, and *J. Clarke* under the *Royal-Exchange* in *Cornhill*, 1723.

Printed in the United States
By Bookmasters